Arduino Biometrics Projects

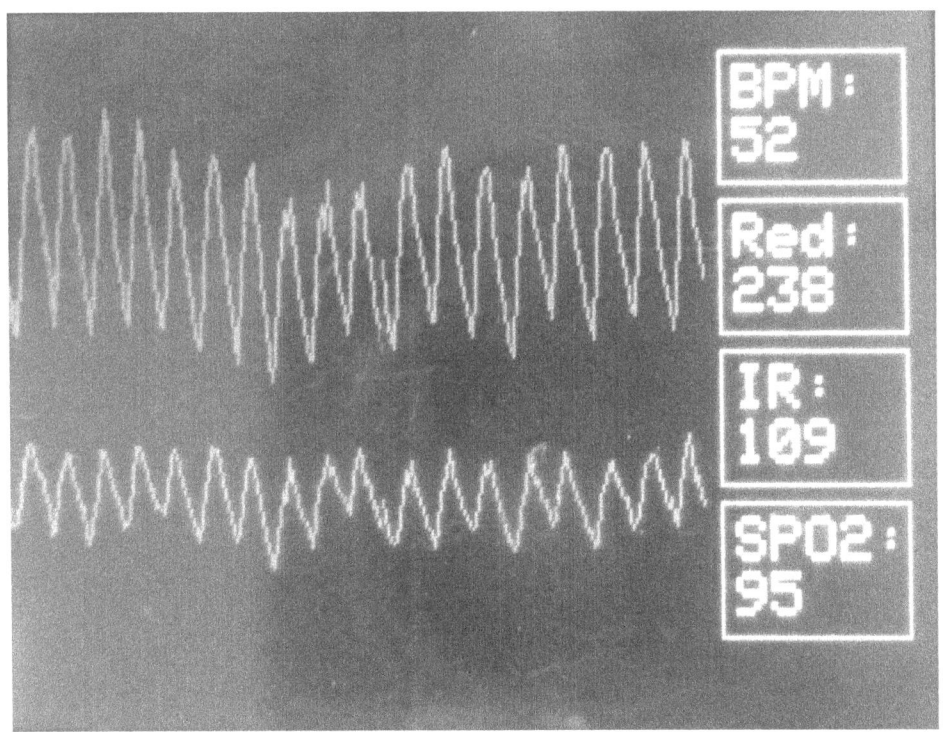

Robert J Davis II

Arduino Biometric Projects, Copyright 2021 by Robert J Davis II

At first I wanted the title of this book to be "Arduino Biologic Projects" as "Bio" and "logic" sounds like a good combination to a techie type of person. However, when you put those two terms together it takes on a different meaning. So "Arduino Biometric Projects" (Measuring biologic parameters) will be the title of this book.

I started writing this book in March 2020 when the Covid 19 pandemic brought everything to a halt. We were almost caught off guard without a blood oxygen sensor at that time. I gave mine to my mom and step-dad because they had an episode where one of their oxygen makers was not working and they did not realize their oxygen levels were low. Then I ordered another one, but it arrived after we had already started to recover from having Covid. Then I thought that it would be a fun project to make my own!

It is now 2020 and it is time to branch out beyond just using the Arduino Uno and Nano. So, some of the projects in this book will be tested with the Arduino UNO and/or the ESP8266 (D1) board. Since the ESP8266 is a 3 volt processor it does not need 1K resistors to interface to 3 volt LCD displays! However, if you are using a five volt UNO or Nano be sure to include those resistors.

We will also be introducing new displays, like the mini OLED displays. These tiny displays do not need a backlight and are easy to hook up. Some OLED's just use only four wires to connect it to the processor.

Besides checking your heartbeat and blood oxygen levels, the sensors covered in this book can do motion detection, IR imaging and can be used for contactless IR temperature detection. Also some environmental sensors are included in this book for monitoring the temperature, humidity and barometric pressure.

As always the projects in this book are meant to be a starting point for your own inventions. Feel free to expand, improve, or personalize these projects. Above all else, have fun!

No guarantee is given or assumed in any way as to the safety or the accuracy of any of these devices. All risk is solely that of the reader / builder. Take normal safety precautions!

Table of Contents

Section 1 - Detecting Heartbeats

1. Heartbeat to 1602 Text LCD Display..................................5
   Heartbeat detector
   1602 16 x 2 LCD

2. Heartbeat to Nokia 84x48 LCD Display.............................11
   Nokia LCD

3. MAX30100/2 to 1602 Text LCD Display....................…...….17
   MAX30100
   MAX30102

4. MAX30102 to Nokia 84x48 LCD Display….....……….......…...23

5. MAX30102 to 1.8 TFT SPI 128 by 160 Color Display…..........29
   1.8 inch TFT SPI Color LCD

6. Max30102 to 2.4 inch LCD Shield.....................................39
   2.4 inch 240x320 Color LCD Shield

7. ECG Display to 1.8 TFT SPI…..……………….….……...…….49
   AD8232

Section 2 - Infrared Projects

9. IR Motion detection to 1602 LCD..…………….….………....55
   HC-SR01 Motion Detector
   FC-51 IR Proximity Switch

10. MLX90614 IR Sensor to 1602 LCD…………………..……. 61
    MLX90614

11. AMG8833 8x8 Thermal Camera and ILI9341 Display…....…65
    AMG8833

Section 3 - Other Sensors/Projects

12. Temperature and Humidity to OLED Display......................71
    DHT11
    96x64 Color OLED
    128x64 White OLED

13. Temperature and Humidity to a Webpage............................81
    D1 Mini

14. Temperature, Pressure and Humidity to OLED..................... 87
    BME280

15. Biometric Watch.. 91
    D1 Mini + SSD1306 + Max30102 + DS1307

16. Colloidal Silver Maker...97

Bibliography..106

Chapter 1

Heartbeat to 1602

Text LCD Display

The simplest way to display someone's heartbeat is with a two line LCD display. To give a graphical presentation some characters of the LCD are reprogrammed to be a bar graph to show the relative amplitude of the heartbeats.

Here is a schematic showing how to wire up the LCD as seen from the top side. The wiring arrangement matches the LCD shield that is sold on eBay. I have soldered the power, ground, and potentiometer on the back of my LCD so that it only needs 8 wires to connect to the Arduino. You can use a four pin header to connect the LCD to the Arduino D4 to D7 data lines to further simplify the wiring. Many of the LCD's used in this book are connected to the Arduino with a header extender to make wiring up the LCD's easier to do.

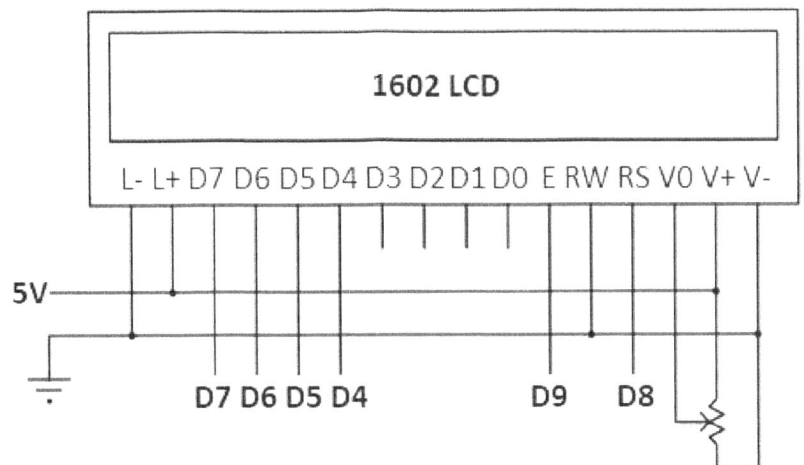

On my heartbeat detector, the green wire is ground, the yellow wire is positive and the orange wire is the heartbeat signal that is connected to A0. The detector consists of an infrared LED, a sensor and an amplifier. The LED lights up your finger and the sensor detects the redness of the blood in your finger that varies with each heartbeat. Now the redness does not vary much and so a lot of amplification is needed. It is a good idea to put some tape on the back side of the detector to protect the exposed electronics. Do not pinch the sensor too tight and give it lots of time to settle down to work properly.

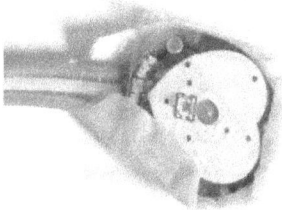

This picture shows the completed heartbeat detector in operation.

This is my program to display heartbeats on a 1602 LCD. It detects the heartbeat, counts the milliseconds between the heartbeats, averages them over five samples and displays the results on the LCD screen.

The program also creates custom characters for a bar graph and then displays the analog value as a bar graph on the bottom half of the LCD screen.

```
// Heartbeat display
// Heartbeat BPM displays on line 1
// Bar Graph of Heartbeat displays on line 2
// By Bob Davis in April 2020

#include <LiquidCrystal.h>
// Variables
int pulsePin = A0; // Pulse Sensor on analog pin 0
int blinkPin = D13; // pin to blink led at each beat
int StartSample = 0;
int EndSample = 0;
int rate[5]; // Array of samples in Milliseconds (MS)
int MS = 0;  // Milliseconds between pulses
int BPM;     // Beats Per Minute
int peak=260;   // Peak voltage
int valley=240; // Minimum voltage
int thresh=250; // Trigger threashold
int sens=80;  // Sensitivity to rise and fall of heartbeat
int Signal; // Incoming raw data from heart sensor
boolean Pulse = false; // "True" when heartbeat detected
int rateTotal = 0;

// LiquidCrystal lcd(13,12,6,5,4,3);//
LiquidCrystal lcd(D8, D9, D4, D5, D6, D7);// LCD shield pinout

// Create graphics lines
byte line1[8]= {B10000, B10000, B10000, B10000, B10000, B10000, B10000, B10000};
byte line2[8]= {B11000, B11000, B11000, B11000, B11000, B11000, B11000, B11000};
```

```
byte line3[8]= {B11100, B11100, B11100, B11100, B11100, B11100,
B11100, B11100};
byte line4[8]= {B11110, B11110, B11110, B11110, B11110, B11110,
B11110, B11110};
byte line5[8]= {B11111, B11111, B11111, B11111, B11111, B11111,
B11111, B11111};

void setup(){
 pinMode(blinkPin,OUTPUT); // pin to blink with heartbeat
 pinMode(pulsePin,INPUT); // Configuring pin A0 as input
 lcd.begin(16,2);  // 16 by 2 LCD
 Serial.begin(9600);
 // Send the custom characters to the LCD.
 lcd.createChar(1, line1);
 lcd.createChar(2, line2);
 lcd.createChar(3, line3);
 lcd.createChar(4, line4);
 lcd.createChar(5, line5);
}

void loop(){
 Signal = analogRead(pulsePin);
 // Display values of BPM Signal on LCD
 lcd.clear();
 lcd.setCursor(0,0);  // First line
 lcd.print("BPM:");
 lcd.print(BPM);
 lcd.setCursor(8,0);  // First line
 lcd.print("MS:");
 lcd.print(MS);
 // Draw Bar Graph of heartbeat
 lcd.setCursor(0,1);  // Second line LCD=Bargraph
 // Draw lower spaces as solid for the bargraph
 int SSignal =Signal/8; // Scale value
 for (int pos=0; pos < SSignal/10; pos++) {lcd.write(5);}
 lcd.setCursor(SSignal/10, 1);
 if (SSignal%10/2 == 0) lcd.write(1);
 if (SSignal%10/2 == 1) lcd.write(2);
 if (SSignal%10/2 == 2) lcd.write(3);
 if (SSignal%10/2 == 3) lcd.write(4);
 if (SSignal%10/2 == 4) lcd.write(5);
```

```
// Find peak, valley and detect change in direction
if (Signal > peak) peak=Signal;  // Find peak
if (Signal < valley) valley=Signal; // Find valley
if (Pulse == false) thresh = (valley+sens); // look for rise
if (Pulse == true) thresh = (peak-sens);    // look for fall

if ((Signal > thresh) && (Pulse == false)){ // Pulse Detected
 Pulse = true; // set Pulse flag
 digitalWrite(blinkPin,HIGH); // turn on pin 13 LED
 }

if ((Signal < thresh) && (Pulse == true)){ // Pulse Finished
 Pulse = false; // reset Pulse flag
 digitalWrite(blinkPin,LOW); // turn off pin 13 LED
 EndSample = millis();
 MS = (EndSample-StartSample);
 StartSample = millis();
 // Reset peak and valley to center
 valley = valley+((peak-valley)/2);
 peak = valley+((peak-valley)/2);
 // BPM = 60000/MS;
 // Keep and average a running total
 rate[5] = rate[4]; // Shift the oldest MS values
 rate[4] = rate[3]; // Shift the oldest MS values
 rate[3] = rate[2]; // Shift the oldest MS values
 rate[2] = rate[1]; // Shift the oldest MS values
 rate[1] = MS; // add the latest MS to array
 rateTotal = (rate[1]+rate[2]+rate[3]+rate[4]+rate[5])/5; // Add up the MS values
 BPM = 60000/rateTotal; // Beats in a minute is BPM

 // display results on computer screen for troubleshooting.
 Serial.print("Beats Per Min=");
 Serial.print("\t");
 Serial.print(BPM);
 Serial.print("\t");
 Serial.print(rate[1]);
 Serial.print("\t");
 Serial.print(rate[2]);
 Serial.print("\t");
```

```
  Serial.print(rate[3]);
  Serial.print("\t");
  Serial.print(rate[4]);
  Serial.print("\t");
  Serial.print(rate[5]);
  Serial.println();
 }
 delay(100); // take a break
}
```

Chapter 2

Heartbeat to Nokia

84x48 LCD Display

I have been working on several Arduino biometric designs for this book. So far in chapter one I have created the two line 1602 LCD display and now we will work with the Nokia 84x48 display. The Nokia display is more fun to work with since I can also do an oscilloscope like display of the heartbeats as they are happening going across the screen. I am working on writing code that works with both an Arduino UNO and with the ESP8266 or the "D1" board.

This is the schematic for connecting the Nokia LCD screen to the Arduino.

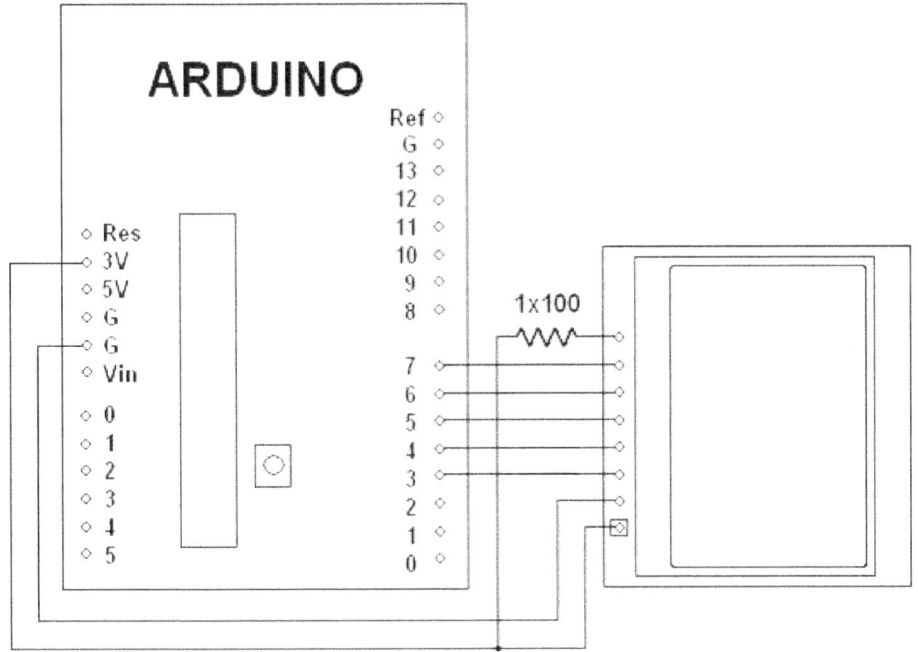

You can connect a NOKIA display to a D1 very easily by using a header extender as they are both three volt devices. For a UNO you should add 1K resistors in series. You only need to connect five pins of the Nokia LCD this way. The other two pins of the LCD are power and ground and they use jumpers to 3.3 Volts and ground. Once again I used a header extender to make connecting the LCD easier to do.

This next picture shows what the connections look like when viewed from above. Note that I have a 100 ohm resistor, to power the LED back light, connected across the two outside pins of the Nokia display. You can also see in the picture that I covered the bottom of the heartbeat sensor with tape to protect it.

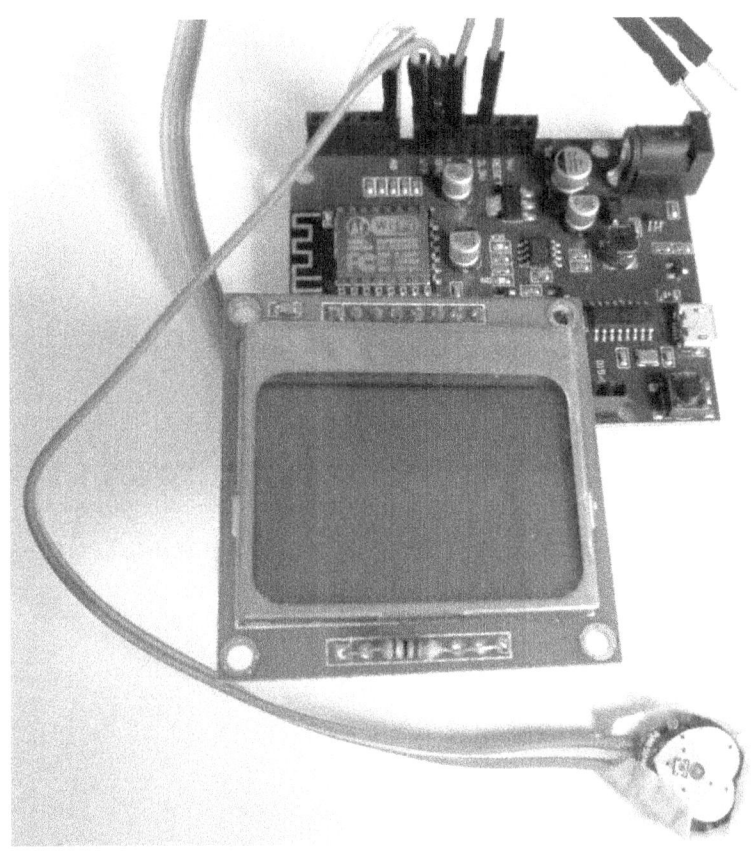

This is a close up picture of the Nokia LCD display showing the beats per minute, milliseconds between beets and the heartbeat pattern.

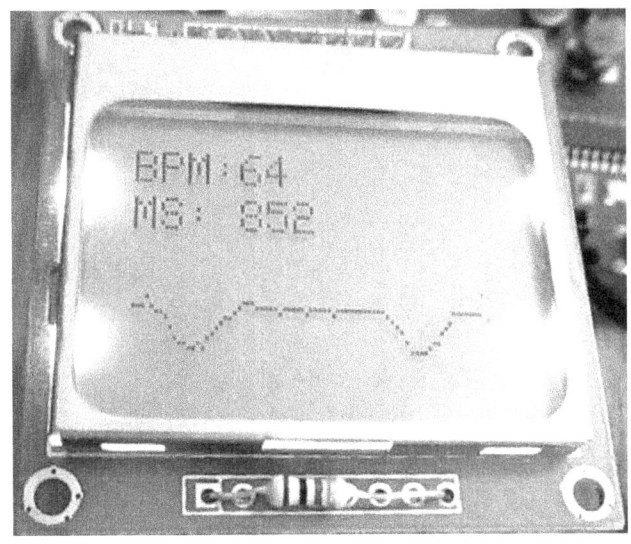

There are lots of code examples for this setup on the internet, but after testing several of them I decided to write my own code.

Here is the code:

```
// NOKIA Heartbeat
// Hearteat BPM displays on line 1
// Scope Trace of Heartbeat displays on lower 1/2
// By Bob Davis in April 2020

#include <SPI.h>
#include <Adafruit_GFX.h>
#include <Adafruit_PCD8544.h>
Adafruit_PCD8544 display = Adafruit_PCD8544(D7, D6, D5, D3, D4);

// Variables
int pulsePin = A0; // Pulse Sensor on analog pin 0
int blinkPin = D13; // pin to blink led at each beat
int StartSample = 0;// Start time MS
int EndSample = 0;  // End time MS
int rate[5];    // Array of samples in Milliseconds (MS)
int MS = 0;    // Milliseconds between pulses
int BPM;      // Beats Per Minute
int peak=800;   // Typical Peak voltage
int valley=500; // Typical Minimum voltage
int thresh=250; // Trigger threshold
int sens=70;   // Sensitivity to rise and fall of heartbeat
int Signal;    // Incoming raw data from heart sensor
int ypos=0;   // Trace Y axis
boolean Pulse = false; // "True" when heartbeat detected
int rateTotal = 0;

void setup(){
  Serial.begin(9600);
  display.begin();
  display.setContrast(50);
  display.clearDisplay();   // clears the screen and buffer
  pinMode(blinkPin,OUTPUT); // pin to blink with heartbeat
  pinMode(pulsePin,INPUT); // Configuring pin A0 as input
}

void loop(){
```

```
Signal = analogRead(pulsePin);
// Display values of BPM Signal on LCD
display.fillRect(0,0,80,20,WHITE); //Clear top
display.setCursor(0,10);  // First line
display.println("MS:");
display.setCursor(24,10);  // First line
display.println(MS);
// Draw the trace of heartbeat
display.drawPixel(ypos,(Signal/10)-40,BLACK);  // Bottom of LCD
ypos=ypos+1;
if (ypos>84) {
  ypos=0;
  display.clearDisplay();  // clears the screen and buffer
}
display.display();  // Update the screen

// Find peak, valley and detect change in direction
if (Signal > peak) peak=Signal;  // Find peak
if (Signal < valley) valley=Signal; // Find valley
if (Pulse == false) thresh = (valley+sens); // look for rise
if (Pulse == true) thresh = (peak-sens);    // look for fall
if ((Signal > thresh) && (Pulse == false)){ // Pulse Detected
  Pulse = true; // set Pulse flag
  digitalWrite(blinkPin,HIGH); // turn on pin 13 LED
}

if ((Signal < thresh) && (Pulse == true)){ // Pulse Finished
  Pulse = false; // reset Pulse flag
  digitalWrite(blinkPin,LOW); // turn off pin 13 LED
  EndSample = millis();
  MS = (EndSample-StartSample);
  StartSample = millis();
  // Reset peak and valley to center
  valley = valley+((peak-valley)/2);
  peak = valley+((peak-valley)/2);
  // BPM = 60000/MS;
  // Keep and average a running total
  rate[5] = rate[4]; // Shift the oldest MS values
  rate[4] = rate[3]; // Shift the oldest MS values
  rate[3] = rate[2]; // Shift the oldest MS values
  rate[2] = rate[1]; // Shift the oldest MS values
```

```
    rate[1] = MS; // add the latest MS to array
    rateTotal = (rate[1]+rate[2]+rate[3]+rate[4]+rate[5])/5; // Add up the
MS values
    BPM = 60000/rateTotal; // Beats in a minute is BPM

    // display results on computer screen for troubleshooting.
    Serial.print("Beats Per Min=");
    Serial.print("\t");
    Serial.print(BPM);
    Serial.print("\t");
    Serial.print(rate[1]);
    Serial.println();
  }
  delay(10); // take a break
}
```

Chapter 3

MAX30100 to 1602

Text LCD Display

For this project we are going to use a MAX30100 to get, not only the pulse rate, but the blood oxygen content as well. The MAX30100 even has a temperature output. Unfortunately the temperature is only readable when the MAX30100 is in a different mode of operation.

The MAX3010X IC's have far less noise, as in a cleaner signal at their outputs. They also have a much faster settling time that the heartbeat detector that was used in the previous projects.

I am working with both the MAX30100 and MAX30102 Blood Oxygen SPO2 Detectors trying to get them to work. First I bought the MAX30100 and then later on I bought the MAX30102 but they simply did not work. I have tried several support libraries to no avail. Then, finally, I came across the issue that the 4.7K pull up resistors are tied to 1.8 volts and they are not high enough of a voltage level for working with a five volt processor.

The first solution to the pull up resistor problem is to remove the three 4.7K resistors (marked 472 in the picture) and replace them with external 4.7K resistors to 3.3 or even to five volts.

The second solution is to cut the run connecting the pull up resistors to 1.8 volts and jumper them over to the 3.3 volt regulator instead. The run to cut is marked in yellow in the next picture; it is located to the right of the top 4.7K resistor. It is cut between the top 472 resistor and the three terminal voltage regulator to its right.

Then add the jumper to 3.3 volts as is marked in red in the picture. The jumper goes from the top of the top 4.7K resistor to the pass through hole located just to the left of the resistor. The through hole is connected to the top right pin of the six pin voltage regulator.

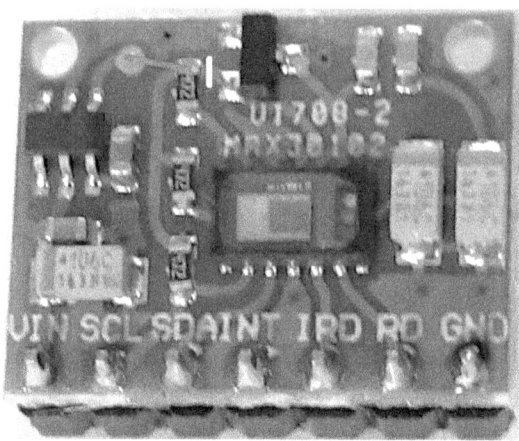

This is a simplified schematic diagram of the circuit board. Basically there are two voltage regulators on the board. The MAX3010X requires both 3.3 and 1.8 volts to operate.

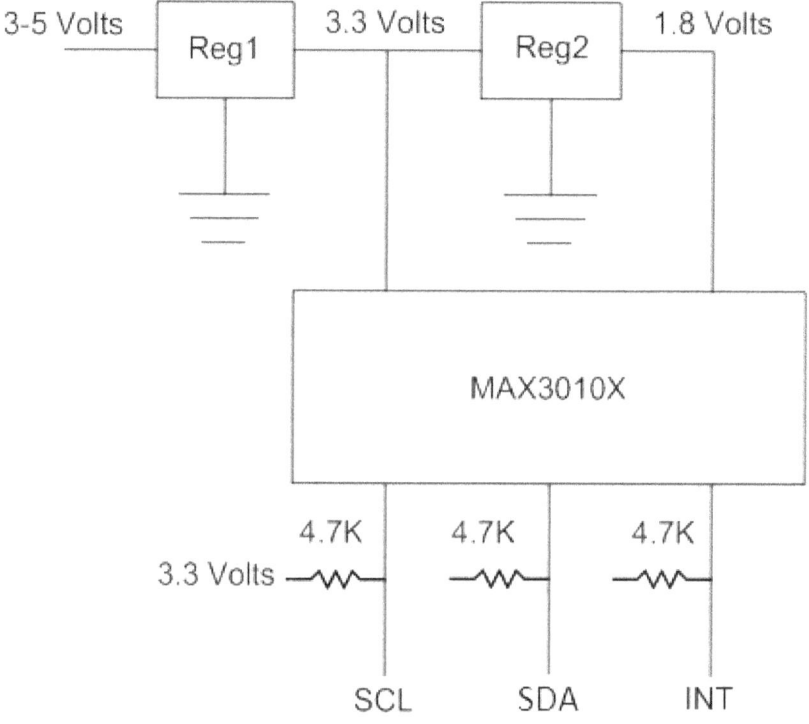

In the schematic Reg1 is a six pin voltage regulator located on the left side of the circuit board. Reg2 is a three pin device located in the top center of the circuit board.

Once the wiring change of connecting the pull up resistors to 3.3 volts is made, the MAX30100 will work great. The "Minimum" example that comes with the library gives beat detection, BPM and SPO2 to the Arduino serial monitor. Just add a little code and redirect it to a LCD and the next picture shows what you get.

The library is called Arduino-MAX30100-master. There are three other example programs included, one gives the raw outputs, and one gives the temperature. Because the examples require that the MAX30100 be in different modes you cannot easily get them to work together.

Only four pins of the MAX3010X are used. Vin and Ground are connected to five volts and ground. SCL and SDA go to SCl and SDA on the Arduino. They are usually the same as A4 and A5 but vary according to the processor type.

The previous picture shows a working blood oxygen level detector made with an Arduino Uno, a 1602 LCD, and a MAX30100.

However the MAX30102 requires a different library and I could not find one nearly as good as the library that is for the MAX30100. Someone commented that, not only is the device ID different (15 instead of 11), but the register addresses are also different.

Well, not only are the register addresses and device ID different, but the register for the LED's changes from one register (4 bits per LED) to two registers (8 bits per LED)! Also some status bits changed between the two devices. I did create a library clone of the MAX30100 to support the MAX30102 as well, it is found on github under bobdavis321.

This code is the "Minimal" MAX30100 example code with the code for the LCD added on.

```
/*
Arduino-MAX30100 oximetry / heart rate integrated sensor library
Copyright (C) 2016  OXullo Intersecans <x@brainrapers.org>
Added LCD by Bob Davis
 */
#include <Wire.h>
#include "MAX30100_PulseOximeter.h"
#include <LiquidCrystal.h>
LiquidCrystal lcd(8, 9, 4, 5, 6, 7);
#define REPORTING_PERIOD_MS     1000

// PulseOximeter is the higher level interface to the sensor
// it offers:
//  * beat detection reporting
//  * heart rate calculation
//  * SpO2 (oxidation level) calculation
PulseOximeter pox;
uint32_t tsLastReport = 0;
// Callback (registered below) fired when a pulse is detected
void onBeatDetected()
{
  Serial.println("Beat!");
  lcd.setCursor(12,0);
```

```
   lcd.print("Beat");
}

void setup()
{
   Serial.begin(9600);
   Serial.print("Initializing..");
   lcd.begin(16,2);
   lcd.setCursor(0,0);
   lcd.print("Initializing...");
   delay(3000);
   lcd.clear();
   // Initialize the PulseOximeter instance
   if (!pox.begin()) {
      Serial.println("FAILED");
      for(;;);
   } else {
      Serial.println("SUCCESS");
   }
   // The default current for the IR LED is 50mA and is changed below
   pox.setIRLedCurrent(MAX30100_LED_CURR_7_6MA);
   // Register a callback for the beat detection
   pox.setOnBeatDetectedCallback(onBeatDetected);
}

void loop()
{
   // Make sure to call update as fast as possible
   pox.update();
   // Asynchronously dump heart rate and oxidation levels to the serial
   // For both, a value of 0 means "invalid"
   if (millis() - tsLastReport > REPORTING_PERIOD_MS) {
      Serial.print("Heart rate:");
      Serial.print(pox.getHeartRate());
      Serial.print("bpm / SpO2:");
      Serial.print(pox.getSpO2());
      Serial.println("%");

      lcd.clear();
      lcd.setCursor(0,0);
      lcd.print("Rate:");
```

```
        lcd.print(pox.getHeartRate());
        lcd.setCursor(0,1);
        lcd.print("SpO2:");
        lcd.print(pox.getSpO2());
        lcd.print("%");

        tsLastReport = millis();
    }
}
```

Chapter 4

MAX30102 to Nokia

84x48 LCD Display

In this chapter we will take the MAX30102 IC to the next level. We will not only display the beats per minute, but we will also graph the heartbeat so that you can see the heartbeats as they are happening on the LCD screen. We covered the Nokia LCD and how to connect it in an earlier chapter of this book.

The heartbeat results from the MAX30102 generally will hover around 75000-100000! A result that is under 50000 indicates that there is likely no finger on the sensor. You can limit the results to 75000 to 110000 to get a better display when you are using the PC based Arduino serial graphing program. You can see the results on the serial plotter as well as the LCD with the supplied program.

The Nokia LCD is also limited to only 48 pixels from the top to the bottom on the screen. The results have to be reduced even more to fit on the screen. To resolve the screen size limit problem, the display has to start at the lowest point in the heartbeat. For that we have to determine the peaks and valleys. Using the valley to start the display on the LCD is the first part of the solution. The peaks would still go completely off the screen. Then the results are divided by 25 to get them into a range that will usually fit on the Nokia screen.

The MAX30105 library comes up with a heart rate that is clearly off in left field. It was usually about two times my actual heart rate! So I wrote my own BPM detector program that is based on the one that I wrote previously in this book. There are fewer variables in this version of my heartbeat detector as some variables were not really needed.

For the LCD heartbeat trace display we need to save the coordinates that the pulse left off at when the last line was drawn for the beginning of a new line that goes from the old point to the new pulse results. So there are several memory values that are saved at the end of the program.

This picture shows the Nokia heartbeat display in operation.

To get a more stable result that does not bounce all over the place we are also going to average the last five sample results together and use that number.

This is the code to make the Nokia heartbeat display work. I suspect there are some included libraries that are not needed as I wrote my own code after the libraries failed to give accurate results. Also the references to the "Y" axis should have been called the "X" axis.

```
/*
  Optical Heart Rate Detection using the MAX30105 Breakout
  Based on: Nathan Seidle @ SparkFun Electronics
  By Bob Davis on 3/29/2021
  This is a demo to show the reading of heart rate or beats per minute
  (BPM) on a Nokia PCD8544 LCD Screen

  Hardware Connections (Break out board to Arduino):
  -5V = 5V (3.3V is allowed)
  -GND = GND
  -SDA = A4 (or SDA)
  -SCL = A5 (or SCL)
  -INT = Not connected
*/

#include <Wire.h>
#include "MAX30105.h"
#include "heartRate.h"
#include "MAX30102_PulseOximeter.h"
#include <Adafruit_GFX.h>
#include <Adafruit_PCD8544.h>
Adafruit_PCD8544 display = Adafruit_PCD8544(7, 6, 5, 3, 4);
MAX30105 particleSensor;

int ypos=0;    // Trace Y axis
int StartSample = 0;
int EndSample = 0;
int rate[5]; // Array of samples in Milliseconds (MS)
int MS = 0;  // Milliseconds between pulses
int BPM;     // Beats Per Minute
long peak=80000;   // initial Peak
long valley=80000; // Minimum valley
long Mvalley=80000; // Memory valley
long redValue=80000;
long irValue=80000;
long MredValue=80000;
long MirValue=80000;
boolean Pulse = false; // "True" when heartbeat detected
boolean Beat = false;
int rateTotal = 0;
```

```
void setup()
{
  Serial.begin(9600);
  Serial.println("Initializing...");
  display.begin();
  display.setContrast(50);
  display.clearDisplay();   // clears the screen and buffer

  // Initialize sensor
  if (!particleSensor.begin(Wire, I2C_SPEED_FAST)) //Use default I2C port, 400kHz speed
  {
    Serial.println("MAX30105 was not found. Please check wiring/power.");
    while (1);
  }

  //Setup to sense a nice looking saw tooth on the plotter
  byte ledBrightness = 0x1F; //Options: 0=Off to 255=50mA
  byte sampleAverage = 8; //Options: 1, 2, 4, 8, 16, 32
  byte ledMode = 3; //Options: 1 = Red only, 2 = Red + IR, 3 = Red + IR + Green
  int sampleRate = 100; //Options: 50, 100, 200, 400, 800, 1000, 1600, 3200
  int pulseWidth = 411; //Options: 69, 118, 215, 411
  int adcRange = 4096; //Options: 2048, 4096, 8192, 16384
  particleSensor.setup(ledBrightness, sampleAverage, ledMode, sampleRate, pulseWidth, adcRange); //Configure

  Serial.print("IR");
  Serial.print('\t');
  Serial.println("RED");
}

void loop() {
  irValue = particleSensor.getIR();
  redValue = particleSensor.getRed();
  // Find peak, valley and detect change in direction
  if (redValue > peak) peak=redValue;  // Find peak
  if (redValue < valley) valley=redValue; // Find valley
  if ((redValue > MredValue) && (Pulse == false)){ // Pulse Detected
```

```
    peak = valley+((peak-valley)/2);
    Pulse = true; // set Pulse flag
    Beat = true;
  }

  if ((redValue < MredValue) && (Pulse == true)){ // Pulse Finished
    Pulse = false; // reset Pulse flag
    EndSample = millis();
    MS = (EndSample-StartSample);
    StartSample = millis();
    // Reset valley to center
    valley = valley+((peak-valley)/2);
    // Keep and average a running total
    rate[5] = rate[4]; // Shift the oldest MS values
    rate[4] = rate[3]; // Shift the oldest MS values
    rate[3] = rate[2]; // Shift the oldest MS values
    rate[2] = rate[1]; // Shift the oldest MS values
    rate[1] = MS; // add the latest MS to array
    rateTotal = (rate[1]+rate[2]+rate[3]+rate[4]+rate[5])/5; // Add up the MS values
    BPM = 60000/rateTotal; // Beats in a minute is BPM
  }

  Serial.print(constrain(irValue, 75000, 100000)); //Send data to plotter
  Serial.print('\t');
  Serial.println(constrain(redValue, 75000, 100000)); //Send data to plotter

  // Draw the trace of heartbeat
  display.drawLine(ypos-1,((MredValue-Mvalley)/25),ypos,((redValue-valley)/25),BLACK);
  ypos=ypos+1;
  if (ypos>84) {
    ypos=0;
    display.clearDisplay();   // clears the screen and buffer
  }
  if (Beat=true){
    display.fillRect(0,40,44,8,WHITE); //Clear box
    display.setCursor(0,40);  // Bottom line
    display.println("BPM: ");
    display.setCursor(24,40);  // Bottom line
    display.println(BPM);
```

```
   Beat=false;
 }
 display.display();  // Update the screen
 MredValue=redValue;
 MirValue=irValue;
 Mvalley=valley;
}
```

Chapter 5

MAX30102 to 1.8 TFT SPI 128 by 160 Color Display

For this project we will use a color LCD to display the live analog Red results in red and the analog Infrared results in blue. Then we can also display text results of BPM and SPO2. This will look even more like the hospital patient status displays.

There appears to be two versions of the 1.8 inch TFT LCD screen. One version has 10 pins and the other version has 16 pins. The pin definitions are written on the bottom of the circuit board so it is just a matter of writing them down before you flip it over and wire it up. We will be using the 16 pin version, but not using the pins that are for a SD memory card.

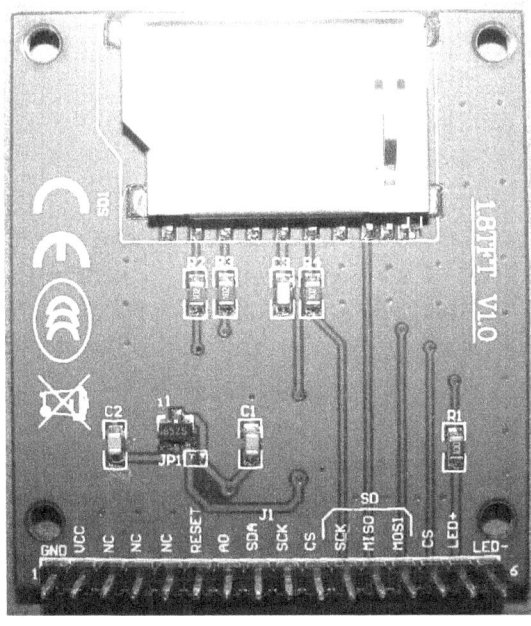

The previous picture is of the back of the 1.8 inch LCD. As you can see the pins are clearly marked. Those under the group called "SD" are for the Secure Digital memory card reader. You could use that to record the results over a long period of time.

The sketches for this LCD use the built in TFT drivers found in version 1.0.5 and above of the Arduino driver. They will not work without this TFT driver being properly configured. This built in TFT driver is the same one as the Adafruit ST7735 driver, it is just renamed as "TFT". There are some Arduino TFT demo programs included, but they are now found under "Retired".

Here is a chart showing the wiring from the Arduino Uno to the TFT LCD screen. I ran jumpers to connect the two grounds and 5V pins together on the back side of the LCD to simplify the wiring.

```
Arduino Uno         1.8 SPI TFT
---------------     --------------
GND                 Pin 01 (GND)
5V (VCC)            Pin 02 (VCC)
Not used            Pin 03 to 05
D8                  Pin 06 (RESET)
D9                  Pin 07 (A0)
D11 (MOSI)          Pin 08 (SDA)
D13 (SCK)           Pin 09 (SCK)
D10 (CS)            Pin 10 (CS)
SD Card             Pins 11 to 14
5V (VCC)            Pin 15 (LED+)
GND                 Pin 16 (LED-)
```

Note that pin one of the LCD is located on the right side as you look at the top of the LCD screen. This next drawing is the schematic diagram showing how to wire the LCD screen up. Unfortunately there is not a 1:1 correspondence between the Arduino and the LCD pins. You will have to find a way to work around the order of the pins.

You can use a four or five pin female header extender plugged into the LCD. The Arduino pins D8 and D9 are connected straight, then bend the next pin up to D11 and double bend the next pin up to D13, then the fifth pin gets a jumper back to pin 10. Then you will need two additional jumpers, one to five volts and one to ground. You can also just wire it up using seven male to female jumper wires.

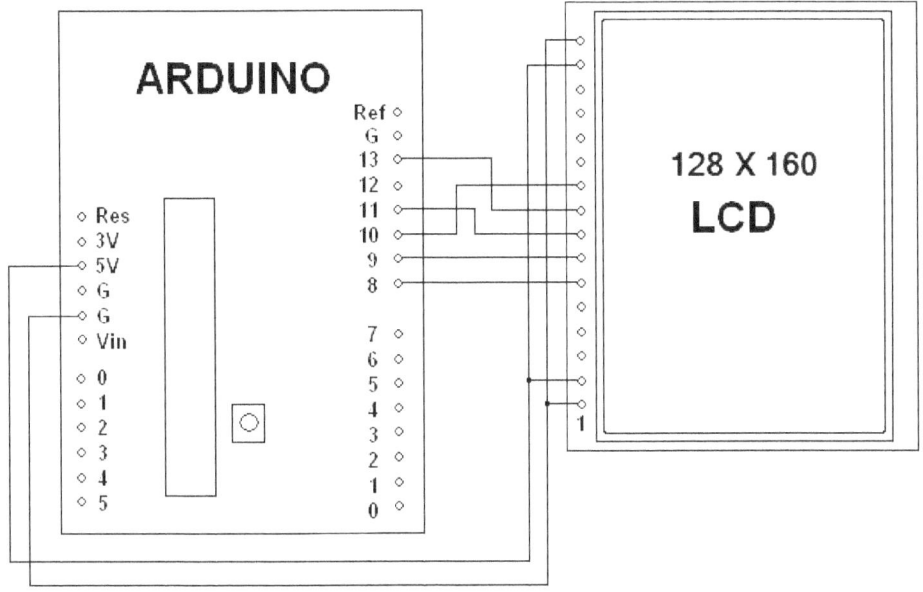

Once again I used my own heartbeat rate detection software. I tried some others and received really strange results. Even with my software the BMP count would sometimes goof up untill I added a second memory sample to compare results to see if the waveform is increasing or decreasing.

This next image is of the Arduino serial plotter results. As you can see the amplitude and center of the signal varies widely. I have seen the analog signal go over 100,000 on occasion. So this signal must be stabilized to fit the results on the smaller LCD screen. To do that we need to detect the valleys or the lowest points and the peaks or the highest points and then average those two and align that to the center of the LCD screen.

Also the results will need to be scaled down to fit it on the small LCD screen. I used a scale factor of 1/15 to shrink the signal amplitude down to size.

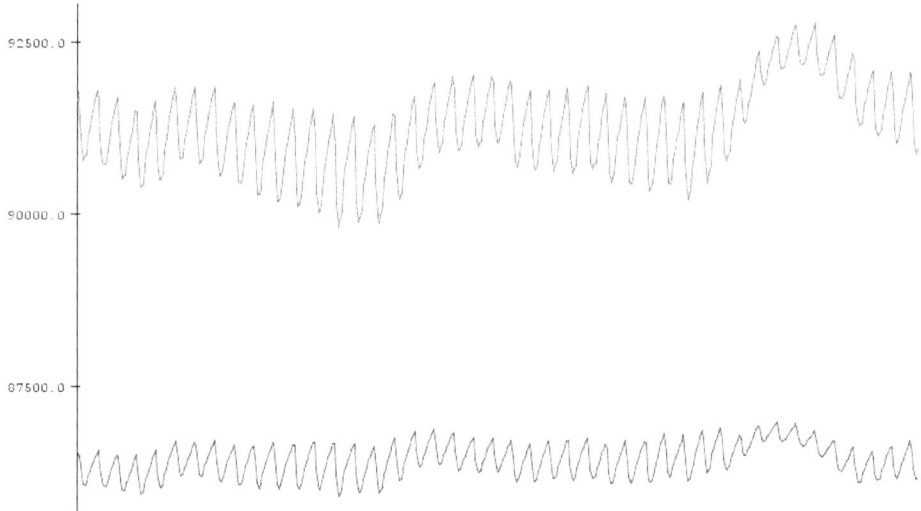

Computing an accurate SPO2 level is very difficult. My program does a very simplified conversion and it is not very accurate. I do not throw out bad results that happen when the peaks and valleys are rapidly shifting. Also the results are supposed to be referenced to a table to find the matching SPO2 level. I just multiplied the reaults by 100.

The formula for SPO2 is Z=AC/DC of red / AC/DC of IR. The AC level is the peak minus the valley and is displayed on the screen. The DC is the average of the waveshape obtained by adding together the samples from one peak to another peak and then dividing that by the number of samples.

Then you are supposed to use a chart to convert the results, Z, to the SPO2 level. Maxim gives a non-chart formula: SpO2 = (-45.06*Z + 30.354)*Z + 94.845 in case you did not want to make a chart of vlaues. At first I just multiplied Z by 100, as Z runs around .9 to 1.0. Later I adjusted the formula to "SPO2R = (-30*Z + 30)*Z + 95". I had to do that to get the same result as a SPO2 detector that I have.

I also made the SPO2 results to be averaged over the last five samples. Even then the SPO2 number that is displayed jumps around a lot. The secret to getting a good reading it to hold yourself and your finger perfectly still.

The next picture shows what you should see on the LCD when the program is running.

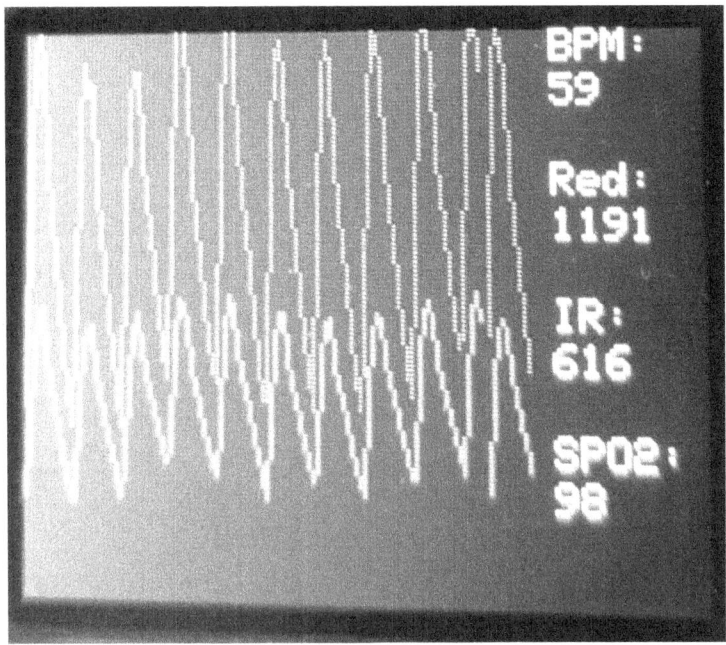

In my opinion the code is too complex and too long with too many variables. I challenge anyone to come up with a simpler solution. Here is my code for your enjoyment.

```
/*
  Optical Heart Rate Detection using the MAX30105 Breakout
  Based on: Nathan Seidle @ SparkFun Electronics
  BY Bob Davis on 3/29/2021
  This is a demo to show the reading of heart rate or beats per minute (BPM) on a SPI 1.8" LCD Screen

  Hardware Connections (MAX30102 to Arduino):
  5V = 5V
  GND = GND
  SDA = A4 (or SDA)
  SCL = A5 (or SCL)
  INT = Not connected
*/

#include <Wire.h>
#include "MAX30105.h"
#include <TFT.h>  // Arduino LCD library
#include <SPI.h>
```

```
// TFT LCD pin definition for the Uno
#define cs   10
#define dc   9
#define rst  8
TFT TFTscreen = TFT(cs, dc, rst);

MAX30105 particleSensor;
// Variables for pulse rate
int StartSample;
int EndSample;
int rate[5]; // Array of samples in Milliseconds (MS)
int rateTotal;
int MS;  // Milliseconds between pulses
int BPM;     // Beats Per Minute
// Peak and valley detection
long peak=80000;      // initial Peak
long valley=80000;    // Minimum valley
long irpeak=80000;    // initial Peak
long irvalley=80000;  // Minimum valley
long redValue;   // Raw Red
long irValue;    // Raw IR
// variables for SPO2
float RedACValue;     // RED AC Component
float IRACValue;      // IR AC Component
float RedACDC;
float IRACDC;
int SPO2R;
int spo2av[5];
int SPO2A;
long SumRed;
long SumIR;
int SampleNo;
// Memory values
long Mvalley=80000;     // Memory valley
long Mirvalley=80000;   // Memory IR valley
long MredValue=80000;   // Memory red value
long MredValue2=80000;  // 2 deep memory
long MirValue=80000;    // Memory IR value
boolean Pulse = false; // "True" when heartbeat detected
// Screen variables
boolean SBeat = false; // Beat detected update screen
```

```
long Sred;    // Average red
long MSred;   // Memory Average red
long Sir;     // Average ir
long MSir;    // Memory Average ir
int xpos=0;   // Trace X axis

void setup()
{
  Serial.begin(9600);
  Serial.println("Initializing...");
  // initialize the display, clear screen
  TFTscreen.begin();
  TFTscreen.background(00, 00, 00);
  // Initialize sensor
  if (!particleSensor.begin(Wire, I2C_SPEED_FAST)) //Use default I2C port, 400kHz speed
  {
    Serial.println("MAX30105 not found. Please check wiring/power. ");
    while (1);
  }
  //Setup to sense a nice looking saw tooth on the plotter
  byte ledBrightness = 0x1F; //Options: 0=Off to FF=50mA
  byte sampleAverage = 8; //Options: 1, 2, 4, 8, 16, 32
  byte ledMode = 3; //Options: 1 = Red only, 2 = Red + IR, 3 = Red + IR + Green
  int sampleRate = 400; //Options: 50, 100, 200, 400, 800, 1000, 1600, 3200
  int pulseWidth = 411; //Options: 69, 118, 215, 411
  int adcRange = 4096; //Options: 2048, 4096, 8192, 16384
  particleSensor.setup(ledBrightness, sampleAverage, ledMode, sampleRate, pulseWidth, adcRange); //Configure
  // Set up serial plotter
  Serial.print("IR");
  Serial.print('\t');
  Serial.println("RED");
}

void loop() {
  irValue = particleSensor.getIR();
  redValue = particleSensor.getRed();
  // Find peak, valley and detect change in direction
```

```
if (redValue > peak) peak=redValue;     // Find peak
if (irValue > irpeak) irpeak=irValue;    // Find peak
if (redValue < valley) valley=redValue; // Find valley
if (irValue < irvalley) irvalley=irValue; // Find ir valley
SumRed=SumRed+redValue-valley;
SumIR=SumIR+irValue-irvalley;
SampleNo++;
if ((redValue > MredValue2) && (Pulse == false)){ // Pulse Detected
  Sred = (peak+valley)/2; // average of peak and valley
  Sir = (irpeak+irvalley)/2; // average of peak and valley

  // Calculate SPO2
  RedACValue = peak-valley;
  IRACValue = irpeak-irvalley;
  RedACDC = RedACValue/(SumRed/SampleNo);//Sum/No= DC
  IRACDC = IRACValue/(SumIR/SampleNo);
  float Z = (RedACDC/IRACDC); // Ratio
  SPO2R = (-30*Z + 30)*Z + 95;
  // Keep and average a running total
  spo2av[5] = spo2av[4]; // Shift the oldest reading
  spo2av[4] = spo2av[3]; // Shift the oldest reading
  spo2av[3] = spo2av[2]; // Shift the oldest reading
  spo2av[2] = spo2av[1]; // Shift the oldest reading
  spo2av[1] = SPO2R; // add the latest reading to array
  SPO2A = ((spo2av[1]+spo2av[2]+spo2av[3]+spo2av[4]+spo2av[5])/5);
// Average
  SumRed=0;
  SumIR=0;
  SampleNo=0;

  peak = (peak+valley)/2; // Reset peak
  irpeak = (irpeak+irvalley)/2; // Reset ir peak
  Pulse = true; // set Pulse flag
  SBeat = true; // Update screen flag
}

if ((redValue < MredValue2) && (Pulse == true)){ // Pulse Finished
  Pulse = false; // reset Pulse flag
  EndSample = millis();
  MS = (EndSample-StartSample);
  StartSample = millis();
```

```
    valley = (peak+valley)/2; // Reset valley
    irvalley = (irpeak+irvalley)/2; // Reset IR valley
    // Keep and average a running total
    rate[5] = rate[4]; // Shift the oldest MS values
    rate[4] = rate[3]; // Shift the oldest MS values
    rate[3] = rate[2]; // Shift the oldest MS values
    rate[2] = rate[1]; // Shift the oldest MS values
    rate[1] = MS; // add the latest MS to array
    rateTotal = ((rate[1]+rate[2]+rate[3]+rate[4]+rate[5])/5); // MS average
    BPM = 60000/rateTotal; // Beats in a minute is BPM
  }

  Serial.print(constrain(irValue, 80000, 120000)); //Send raw data to plotter
  Serial.print('\t');
  Serial.println(constrain(redValue, 80000, 120000)); //Send raw data to plotter

  // Clear Ahead then Draw the trace of heartbeat
  TFTscreen.stroke(0,0,0);   // Black
  TFTscreen.rect(xpos+1,0,1,128);  // Clear ahead

  TFTscreen.stroke(00, 00, 250);  // Red
  TFTscreen.line(xpos-1,(((MredValue-MSred)/15)+40),xpos,(((redValue-Sred)/15)+40));
  TFTscreen.stroke(250, 00, 00);  // Blue
  TFTscreen.line(xpos-1,(((MirValue-MSir)/15)+80),xpos,(((irValue-Sir)/15)+80));

  xpos=xpos+1;
  if (xpos>120) {
    xpos=0;
    TFTscreen.stroke(0,0,0);    // Black
    TFTscreen.rect(0,0,1,128);  // Clear left side
  }
  if (SBeat==true){
    TFTscreen.stroke(0,0,0);   // Black
    TFTscreen.fill(0,0,0);
    TFTscreen.rect(120,0,60,128);  // Clear right side
    TFTscreen.stroke(250, 250, 250);  // White
```

```
TFTscreen.setCursor(125,00);  // Right Side
TFTscreen.println("BPM:");
TFTscreen.setCursor(125,10);
TFTscreen.println(BPM);
TFTscreen.setCursor(125,30);
TFTscreen.println("Red:");
TFTscreen.setCursor(125,40);
TFTscreen.println(RedACValue,0);
TFTscreen.setCursor(125,60);
TFTscreen.println("IR:");
TFTscreen.setCursor(125,70);
TFTscreen.println(IRACValue,0);
TFTscreen.setCursor(125,90);
TFTscreen.println("SPO2:");
TFTscreen.setCursor(125,100);
TFTscreen.println(SPO2A);

  SBeat=false;
}
MredValue2=MredValue;
MredValue=redValue;
MirValue=irValue;
Mvalley=valley;
Mirvalley=irvalley;
MSred=Sred;
MSir=Sir;
}
```

Chapter 7

MAX30102 to 2.4 inch LCD Shield

Using a bigger screen will result in a much better display of the waveform and data! There is a 2.4 inch 240 by 320 pixel color LCD Shield that is commonly available on eBay and Amazon. There is also a 3.2 and a 3.5 inch LCD shield. However, these shields use almost all of the Arduino pins, and do not leave a single pin accessible to you. As a minimum, we will need A4 and A5 because they are also known as SCL and SDA and are used to connect the MAX30102 to the Arduino.

The screen that measures 2.4 inches offers a resolution of 320 by 240. That is the default size and is supported by the Adafruit LCD drivers. The screen that measures 3.2 inches diagonal has a resolution of 400 horizontal pixels by 240 vertical pixels. The largest screen is 3.5 inches and has a resolution of 480 by 320 pixels. Using one of these screens will result in lots of available pixels allowing more data and finer graphs.

The 2.4 and the 3.5 inch LCD screens use the same pin out as can be seen in the previous picture. But the 3.2 inch LCD rearranges some of the pins as compared to the 2.4 or 3.5 inch models. For that reason I recommend that you avoid using the 3.2 inch LCD shield. Otherwise you will have to remap the data bits in the driver software.

This chart shows the pin assignments for the three sizes of LCD's:

LCD Shield Pin Name	2.4 LCD 3.5 LCD	3.2 LCD
GND	GND	GND
5V	5V	5V
CS	A3	A3
RS	A2	A2
WR	A1	A1
RD	A0	A0
RST	A4	RESET
LED	GND	GND
DB0	D8	D8
DB1	D9	D9
DB2	D2	D10
DB3	D3	D11
DB4	D4	D4
DB7	D5	D13
DB6	D6	D6
DB7	D7	D7

Note that A4 is used for the LCD "Reset" signal on the 2.4 and the 3.5-inch screens. That will have to be routed to the Arduino Reset instead to free up that pin for analog inputs or to use SCL and SDA. Also, if the temperature sensing IC that is called "U3" in the picture above is installed it uses A5 as its CS (Chip Select). I am not sure as to how that would affect my software as none of the LCD shields that I have, have this IC installed.

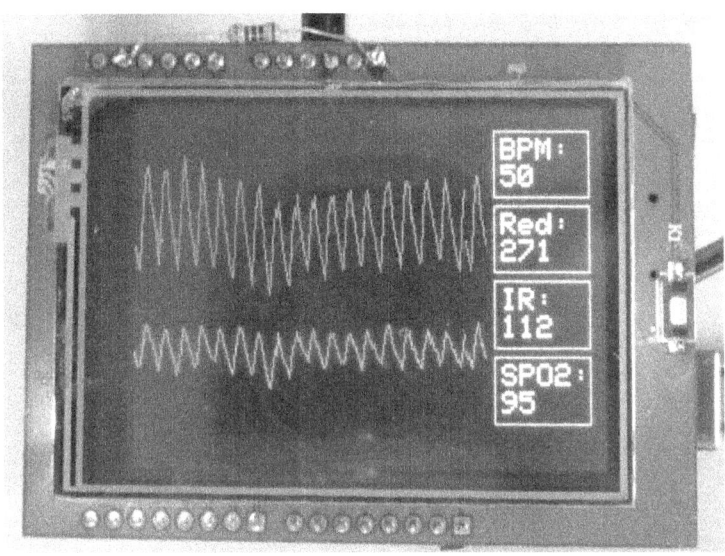

To find out what pins are actually needed, I used header extenders to test disconnecting the unused pins and verify that the LCD shield still works. On one side of the Arduino pins D2 through D9 are used as D0 to D7 for the LCD. This can be seen in the next picture. The other pins on this side of the Arduino are used for the memory card slot and they are not used for any of the projects found in this book.

On the other side of the Arduino shield we need A0 to A4, GND, 5V and Reset. As mentioned earlier, we need A4 for the MAX30102. To free up A4 you can connect a 100 ohm resistor from the LCD shield A4 to Reset. Then you can disconnect A4 and everything should still work just fine. This modification can be seen in the next picture.

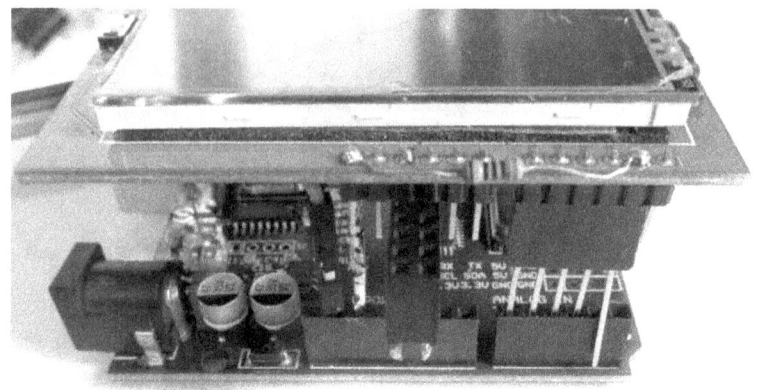

Now you can connect the MAX30102 just as was done in the previous projects.

Here is the code modified for the much higher resolution screen.

```
/*
 BY Bob Davis on 3/29/2021
 This is a demo to show the reading of heart rate and SPO2 on a 2.4" LCD Screen

 Hardware Connections (MAX30102 to Arduino):
 5V = 5V
 GND = GND
 SDA = A4 (or SDA)
 SCL = A5 (or SCL)
 INT = Not connected

Use these 8 data lines to connect to the LCD:
For the Arduino Uno, Duemilanove, Diecimila, etc.:
  D0 connects to digital pin 8  (Notice these are
  D1 connects to digital pin 9   NOT in order!)
  D2 connects to digital pin 2
  D3 connects to digital pin 3
  D4 connects to digital pin 4
  D5 connects to digital pin 5
  D6 connects to digital pin 6
  D7 connects to digital pin 7

The control pins for the LCD can be assigned to any digital or
analog pins...but we'll use the analog pins.
```

```
*/
#include <Wire.h>
#include "MAX30105.h"
#include <SPI.h>

#include <Adafruit_GFX.h>    // Core graphics library
#include <Adafruit_TFTLCD.h> // Hardware-specific library
// TFT LCD pin definition for the Uno
#define LCD_RESET A7 // FAKE-Connect LCD Reset to Arduino's reset
#define LCD_CS A3 // Chip Select goes to Analog 3
#define LCD_CD A2 // Command/Data goes to Analog 2
#define LCD_WR A1 // LCD Write goes to Analog 1
#define LCD_RD A0 // LCD Read goes to Analog 0

// Assign human-readable names to some common 16-bit color values:
#define BLACK   0x0000
#define BLUE    0x001F
#define RED     0xF800
#define GREEN   0x07E0
#define CYAN    0x07FF
#define MAGENTA 0xF81F
#define YELLOW  0xFFE0
#define WHITE   0xFFFF
Adafruit_TFTLCD tft(LCD_CS, LCD_CD, LCD_WR, LCD_RD, LCD_RESET);

MAX30105 particleSensor;
// Variables for pulse rate
int StartSample;
int EndSample;
int rate[5]; // Array of samples in Milliseconds (MS)
int rateTotal;
int MS;  // Milliseconds between pulses
int BPM;     // Beats Per Minute
// Peak and valley detection
long peak=80000;       // initial Peak
long valley=80000;     // Minimum valley
long irpeak=80000;     // initial Peak
long irvalley=80000;   // Minimum valley
long redValue;   // Raw Red
long irValue;    // Raw IR
```

```
// variables for SPO2
float RedACValue;    // RED AC Component
float IRACValue;     // IR AC Component
float RedACDC;
float IRACDC;
int SPO2R;
int spo2av[5];
int SPO2A;
long SumRed;
long SumIR;
int SampleNo;
// Memory values
long Mvalley=80000;     // Memory valley
long Mirvalley=80000;   // Memory IR valley
long MredValue=80000;   // Memory red value
long MredValue2=80000;  // 2 deep memory
long MirValue=80000;    // Memory IR value
boolean Pulse = false; // "True" when heartbeat detected
// Screen variables
boolean SBeat = false; // Beat detected update screen
long Sred;    // Average red
long MSred;   // Memory Average red
long Sir;     // Average ir
long MSir;    // Memory Average ir
int xpos=0;   // Trace X axis

void setup()
{
  Serial.begin(9600);
  Serial.println("Initializing...");
  // initialize the display, clear screen
  uint16_t identifier = 0x9341;
  // Options: 0x8357 0x7575 0x7575 0x9328 0x9325
//  tft.reset();
  tft.begin(identifier);
  tft.fillScreen(BLACK);
  tft.setRotation(1);

  // Initialize sensor
  if (!particleSensor.begin(Wire, I2C_SPEED_FAST)) //Use default I2C port, 400kHz speed
```

```
  {
    Serial.println("MAX30105 was not found. Please check wiring/power. ");
    while (1);
  }
  //Setup to sense a nice looking saw tooth on the plotter
  byte ledBrightness = 0x1F; //Options: 0=Off to FF=50mA
  byte sampleAverage = 8; //Options: 1, 2, 4, 8, 16, 32
  byte ledMode = 3; //Options: 1 = Red only, 2 = Red + IR, 3 = Red + IR + Green
  int sampleRate = 400; //Options: 50, 100, 200, 400, 800, 1000, 1600, 3200
  int pulseWidth = 411; //Options: 69, 118, 215, 411
  int adcRange = 4096; //Options: 2048, 4096, 8192, 16384
  particleSensor.setup(ledBrightness, sampleAverage, ledMode, sampleRate, pulseWidth, adcRange); //Configure
  // Set up serial plotter
  Serial.print("IR");
  Serial.print('\t');
  Serial.println("RED");
}

void loop() {
  irValue = particleSensor.getIR();
  redValue = particleSensor.getRed();
  // Find peak, valley and detect change in direction
  if (redValue > peak) peak=redValue;     // Find peak
  if (irValue > irpeak) irpeak=irValue;    // Find peak
  if (redValue < valley) valley=redValue; // Find valley
  if (irValue < irvalley) irvalley=irValue; // Find ir valley
  SumRed=SumRed+redValue-valley;
  SumIR=SumIR+irValue-irvalley;
  SampleNo++;
  if ((redValue > MredValue2) && (Pulse == false)){ // Pulse Detected
    Sred = (peak+valley)/2; // average of peak and valley
    Sir = (irpeak+irvalley)/2; // average of peak and valley

    // Calculate SPO2
    RedACValue = peak-valley;
    IRACValue = irpeak-irvalley;
```

```
    RedACDC = RedACValue/(SumRed/SampleNo); //Sum/No=average or DC
    IRACDC = IRACValue/(SumIR/SampleNo);
    float Z = (RedACDC/IRACDC); // Ratio
    SPO2R = (-30*Z + 30)*Z + 95; // Modified formula

    // Keep and average a running total
    spo2av[5] = spo2av[4]; // Shift the oldest reading
    spo2av[4] = spo2av[3]; // Shift the oldest reading
    spo2av[3] = spo2av[2]; // Shift the oldest reading
    spo2av[2] = spo2av[1]; // Shift the oldest reading
    spo2av[1] = SPO2R; // add the latest reading to array
    SPO2A = ((spo2av[1]+spo2av[2]+spo2av[3]+spo2av[4]+spo2av[5])/5); // Average
    SumRed=0;
    SumIR=0;
    SampleNo=0;

    peak = (peak+valley)/2; // Reset peak
    irpeak = (irpeak+irvalley)/2; // Reset ir peak
    Pulse = true; // set Pulse flag
    SBeat = true; // Update screen flag
  }

  if ((redValue < MredValue2) && (Pulse == true)){ // Pulse Finished
    Pulse = false; // reset Pulse flag
    EndSample = millis();
    MS = (EndSample-StartSample);
    StartSample = millis();

    valley = (peak+valley)/2; // Reset valley
    irvalley = (irpeak+irvalley)/2; // Reset IR valley
    // Keep and average a running total
    rate[5] = rate[4]; // Shift the oldest MS values
    rate[4] = rate[3]; // Shift the oldest MS values
    rate[3] = rate[2]; // Shift the oldest MS values
    rate[2] = rate[1]; // Shift the oldest MS values
    rate[1] = MS; // add the latest MS to array
    rateTotal = ((rate[1]+rate[2]+rate[3]+rate[4]+rate[5])/5); // Add up the MS values
    BPM = 60000/rateTotal; // Beats in a minute is BPM
```

```
  }

  Serial.print(constrain(irValue, 80000, 120000)); //Send raw data to plotter
  Serial.print('\t');
  Serial.println(constrain(redValue, 80000, 120000)); //Send raw data to plotter

  // Clear Ahead then Draw the trace of heartbeat
  tft.drawRect(xpos,0,1,240,BLACK);  // Clear ahead
  tft.drawLine(xpos-1,(((MredValue-MSred)/5)+80),xpos,(((redValue-Sred)/5)+80),RED);
  tft.drawLine(xpos-1,(((MirValue-MSir)/5)+160),xpos,(((irValue-Sir)/5)+160),BLUE);

  xpos=xpos+1;
  if (xpos>239) {
    xpos=0;
  }
  if (SBeat==true){

//    TFTscreen.fill(0,0,0);
    tft.setTextSize(2);
    tft.fillRect(240,0,80,240,BLACK);  // Clear right side
    tft.setTextColor(WHITE);
    tft.drawRect(245,15,65,45,GREEN);  // Clear right side
    tft.drawRect(245,65,65,45,GREEN);  // Clear right side
    tft.drawRect(245,115,65,45,GREEN);  // Clear right side
    tft.drawRect(245,165,65,45,GREEN);  // Clear right side
    tft.setCursor(250,20);  // Right Side
    tft.println("BPM:");
    tft.setCursor(250,38);
    tft.println(BPM);
    tft.setCursor(250,70);
    tft.println("Red:");
    tft.setCursor(250,88);
    tft.println(RedACValue,0);
    tft.setCursor(250,120);
    tft.println("IR:");
    tft.setCursor(250,138);
    tft.println(IRACValue,0);
```

```
    tft.setCursor(250,170);
    tft.println("SPO2:");
    tft.setCursor(250,188);
    tft.println(SPO2A);

    SBeat=false;
  }
  MredValue2=MredValue;
  MredValue=redValue;
  MirValue=irValue;
  Mvalley=valley;
  Mirvalley=irvalley;
  MSred=Sred;
  MSir=Sir;
}
```

Chapter 8

ECG Display to 1.8 TFT SPI

In this chapter we will use a device that uses electrodes that attach to you. All safety precautions should be made as this can be a dangerous thing to do. We will be using the AD8232 IC, it is called a "Single-Lead Heart Rate Monitor". In actuality it requires three connections to give a good response. This type of device requires lots of amplification to work so there is a tendency to have some noise in the output. This noise makes it harder to detect the heartbeats.

Here is a block diagram of the AD8232. It can detect if the leads are disconnected and has internal switches to attempt to stabilize the output.

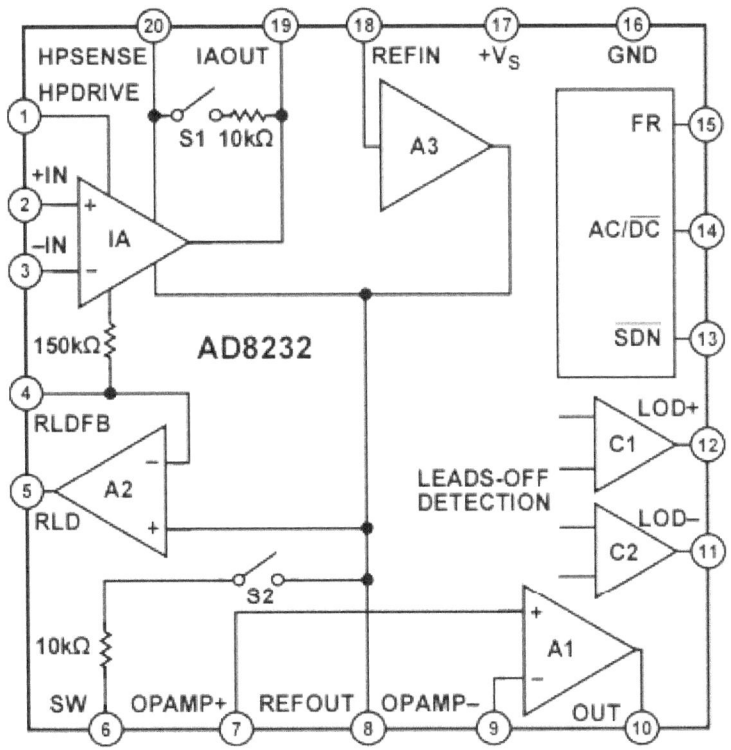

The AD8232 usually comes with disposable electrodes to attach to the person being monitored. However you can also purchase reusable or "Passive" electrodes that can be cleaned and reused. Here is a picture of the ad for the passive electrodes on eBay.

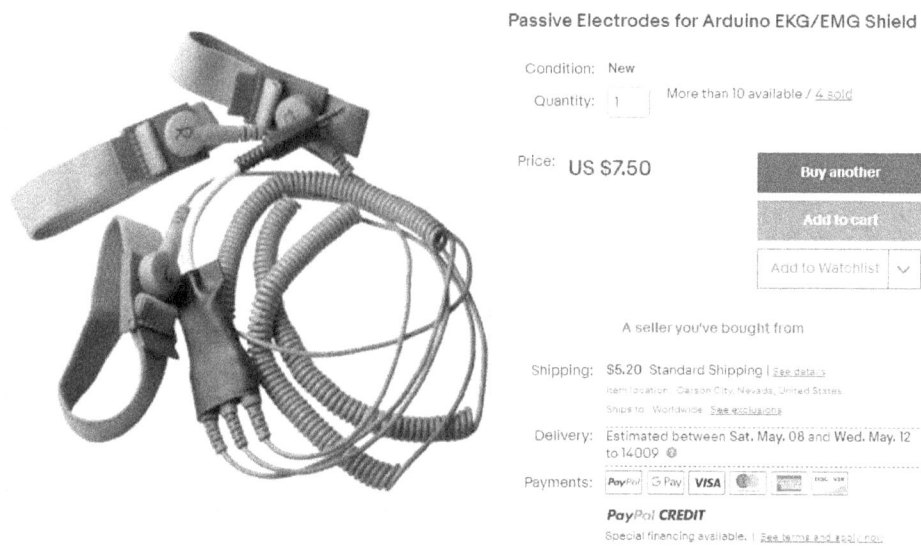

In my limited experience the passive electrodes give more noise, likely because of the longer leads. The electrodes that attach to you give less noise but their wires are way too short! The attached electrodes are color coded. Red is for right arm, yellow is for the left arm and green is for the right leg. The passive electrodes are marked "R" for right "L" for left and "D" for the leg. You want to place the electrodes as close to the body as possible, not way down on the arms or legs, as doing that will give more noise..

The pins of the AD8232 are connected as follows:

```
AD8232          Arduino
--------        -------
Gnd             Gnd
3.3V            3.3V
Output          A0
LD-             D6 - This is output out of range because its disconnected
LD+             D7 - This is output out of range because its disconnected
SDN             NC - This is shut down and is not used.
```

Coming up next is a picture of the AD8232 circuit board with the Arduino pin connections written into the picture on the left side. Most Arduino programs use D10 and D11 for the "Leads Off" (LO) signals. I changed that to use D6 and D7 instead to make it easier to work with the LCD screen connections. When a "Leads Off" condition is detected the Arduino will stop taking samples because it is only getting garbage.

GND
3.3V
A0
D6
D7
NC

This next picture shows what the results looked like on the Arduino serial plotter. I noticed that the results on the plotter are actually upside down. The "spikes" should be pointed upwards. The higher noise level is from using the "passive" electrodes with their longer lead lengths. Perhaps moving the leads further away from all electronics would reduce that noise.

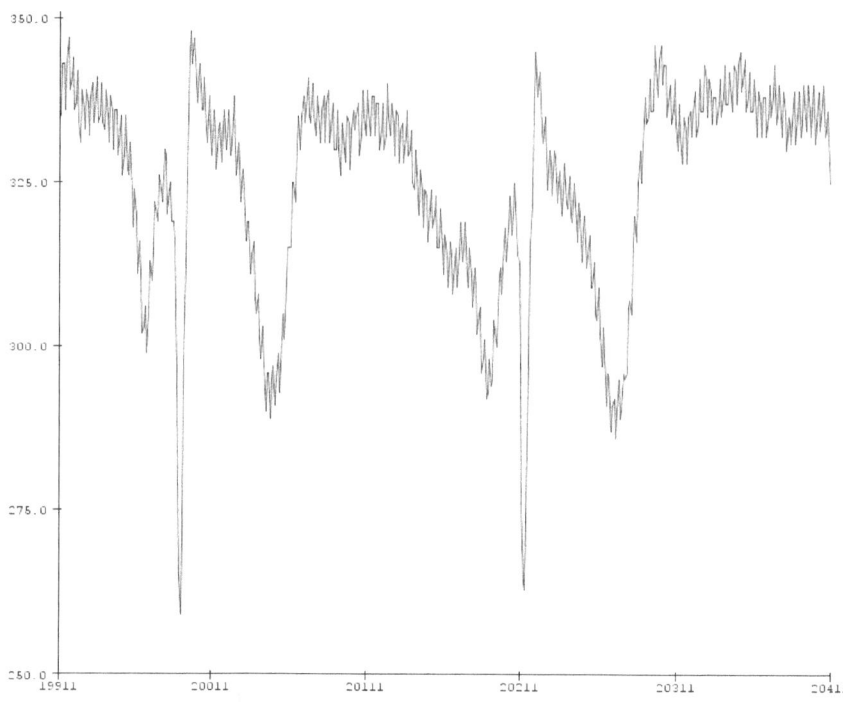

This picture is of the results shown on the small LCD screen. This time it is right side up, but backwards right to left. The amplitude is divided by two so that it does not go off the top and bottom of the screen as much.

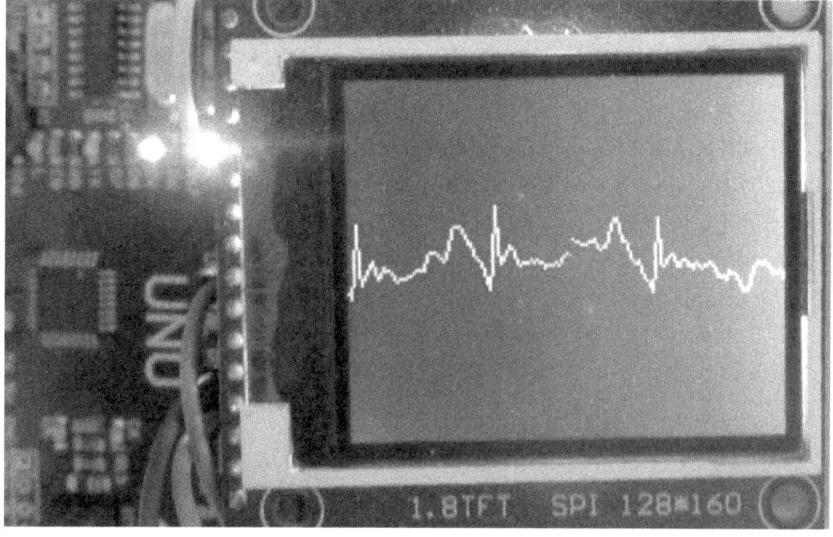

This is my code, there is no heartbeat rate detection routine because of the all the noise in the signal. I guess if "rate" and "threshold" variables were used to ignore the smaller pulses and the repeat pulses then the heartbeat rate could possibly then be accurately detected.

```
// AD8232 to 1.8 SPI LCD
// 5/5/2021 by Bob Davis

#include <Wire.h>
#include <TFT.h>  // Arduino LCD library
#include <SPI.h>

// TFT LCD pin definition for the Uno
#define cs   10
#define dc   9
#define rst  8
TFT TFTscreen = TFT(cs, dc, rst);
int xpos;
int lanalog;
int manalog;

void setup() {
  // initialize the serial communication:
  Serial.begin(9600);
  pinMode(6, INPUT); // Setup for leads off detection LO +
  pinMode(7, INPUT); // Setup for leads off detection LO -
  TFTscreen.begin();
  TFTscreen.background(00, 00, 00);
}

void loop() {
  // Leads off detection
  if((digitalRead(6) == 1)||(digitalRead(7) == 1)){
    Serial.println('!');
  }
  else{
  // send the value of analog input 0:
    lanalog=analogRead(A0);
    Serial.println(lanalog);
  }
  xpos=xpos+1;
  if (xpos>160) {
    xpos=0;
```

```
    }
    TFTscreen.stroke(0,0,0);    // Black
    TFTscreen.line(xpos,0,xpos,128);  // Clear ahead
    TFTscreen.stroke(00,00,250);    // Red
    TFTscreen.line(xpos-1,(manalog/2)-100,xpos,(lanalog/2)-100);
    manalog=lanalog;

    //Wait for a bit to keep serial data from saturating
    delay(20);
}
```

Chapter 9

IR Motion Detection

To 1602 LCD

In this next section of this book we will cover Infrared (IR) motion, temperature detection, and even IR video. Infrared is light that is just outside of the visible light spectrum. It is given off by almost any warm object. We can use it to detect the presence of people or animals as well as to take their temperature and even to look for "infected" people that have a fever. It can also be used to examine houses and buildings for heat leakages.

We will start with simple IR devices and work our way up to more complex devices. There are two devices we will cover in this chapter. The first device is IR reflection detection sometimes called the "FC-51" and the second device is IR motion detection called "HC-SR501".

The IR reflection device works by transmitting pulses of IR light and then looking for those pulses to be reflected back in an IR receiver. These devices work as proximity switches, they trip when something is near them. Their range is limited to just an inch or two from the transmitter. Also, their range can be affected by the reflectivity of the detected surface. These devices vary widely in their appearance, but a typical one that is commonly found on eBay is pictured below.

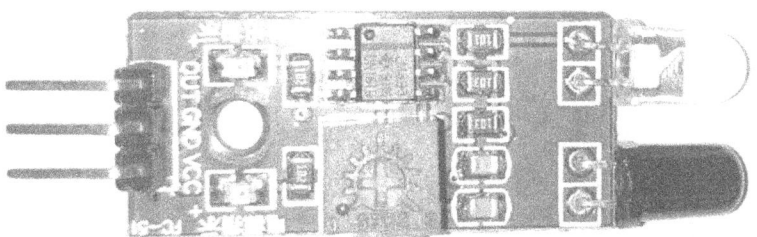

The IR motion detection devices are more complex. They have two IR sensors that cover two regions of sensitivity. When something that is giving off IR light, passes from one sensor region to another sensor region, then "motion" is detected. These sensors are usually covered by special IR lenses that allow them to detect motion in more than one direction.

The actual sensor device looks like a TO-5 transistor with a window on top. These sensors have a sensitivity range that covers from a few inches to many feet! This next picture shows what one looks like from above showing the special lens that covers the sensor.

This next picture shows what the IR motion detetor looks like with its cover removed so You can see the sensor device.

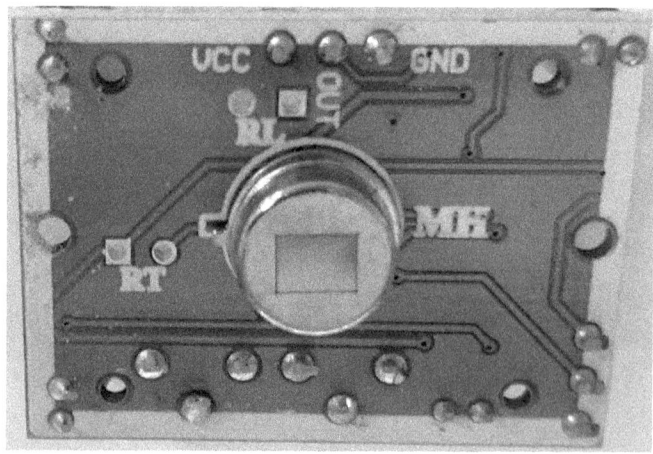

The next picture, shows a typical HC-SR501 from the bottom side so you can see the jumpers and trimmer resistors. It has a jumber that can determine if it is triggered just once or it can be triggered several times. There are two variable resistors for andjusting the trigger time and the sensitivity. Some models do not have the trigger jumper on them. You can turn the delay all the way down and essentially get the same effect as the repeat trigger function.

Several images, that are found on the internet, have the three connection pins labeled wrong in the picture! Fortunatly there is a diode that should protect the device from being damaged by being connected up incorectly. So, be sure to check your connections, that they match the labels on the board before powering one up.

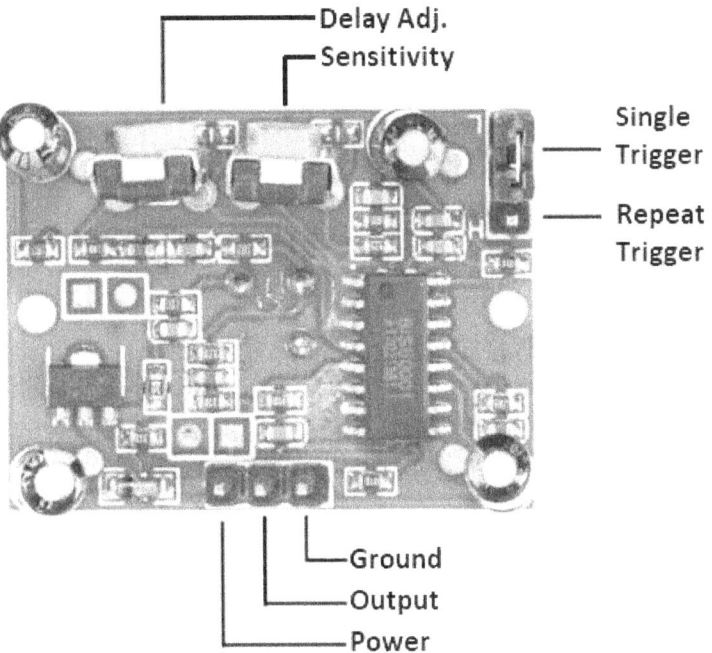

In operation there were so many false triggers with the IR motion detectors that the only reliable results were from using the proximity detectors. That test setup can be seen in the next picture.

Here is the code for the IR motion detector. An arrow appears on the LCD screen to indicate the direction of the motion as you wave your hand over the IR detectors.

If you want to use this code with the IR motion detectors you will need to reverse the "less-than" to a "greater-than" in "IN1 < trigger" and in "IN0 < trigger". The IR proximity switches trigger output is reversed in polarity from the motion detectors output.

```
/*
 * IR Motion detection demonstration
 * By Bob Davis
 * Detects motion right to left or left to right
 * Draws an arrow on the LCD screen indicating the results
 */

#include <LiquidCrystal.h>
LiquidCrystal lcd(8, 9, 4, 5, 6, 7);
int Delay=10;
int trigger=500;
int left=0;
int right=0;
```

```
void setup() {
  Serial.begin(9600);
  lcd.begin(16, 2);
}

void loop() {
  lcd.clear();
  // Read the input values greater than 512 is a positive value
  int IN0 = analogRead(A0);
  int IN1 = analogRead(A1);

  // Check first detector
  if (IN1 < trigger){
    left=1;
    Delay=5;
    if (right==1){
      lcd.setCursor(10, 0);
      lcd.print("<----");
    }
  }

  // Check second detector
  if (IN0 < trigger){
    right=1;
    Delay=5;
    if (left==1){
      lcd.setCursor(10, 0);
      lcd.print("---->");
    }
  }

  Delay--;
  if (Delay == 0){
    left=0;
    right=0;
    lcd.clear();
  }

  // Display the results
  lcd.setCursor(0, 0);
```

```
    lcd.print("A0 ");
    lcd.print(IN0);
    lcd.setCursor(0, 1);
    lcd.print("A1 ");
    lcd.print(IN1);

    delay(500);
}
```

Chapter 10

MLX90614 IR Sensor

To 1602 LCD

For this next project we are going to introduce the MLX90614 IR sensor. This sensor is usually used to measure the IR output from a single spot and then display that information on a screen. It is a low-cost non-contact thermometer on a chip. It can be used to test the temperature of a person fairly quickly as they pass by. It works best at close range like around one inch away.

Some manufacturers add a laser pointer to make sure that you are pointing at the person's forehead, as that gets the best temperature results. The MLX90614 has two temperature sensors inside the IC. One temperature sensor is for the Ambient or chip temperature, the other temperature sensor is the IR or object temperature.

The MLX914 only has four pins. It is housed in a TO39 package, or basically a small round metal transistor style package. You can purchase one that is mounted on a small circuit board or just the bare part and then attach four wires to it like I did in the picture below.

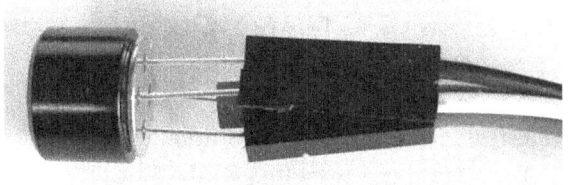

The MLX90614 IC comes in either a 3-5 volt version or in a 5-7 volt version. The manufacturer recommends that you add two pull up resistors when it is interfaced to the SMBus as can be seen in the following diagram on the next page. I tested it without the pull up resistors and it worked fine that way. The drawing below shows the MLX9061 IC as seen from above.

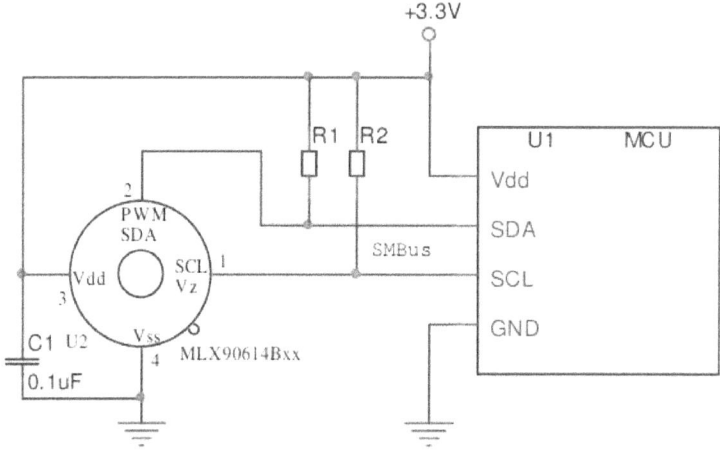

This drawing is the MLX90614 connections as seen from below. If you do not purchase it mounted on a circuit board this is the diagram you will need to connect to it.

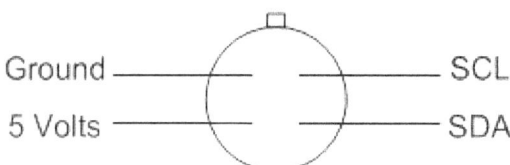

I installed the adafruit "Adafruit-MLX90614-Library-master" library. Then I ran the one example program that is included with that library. It worked perfectly, displaying the results on the attached computer with the Arduino serial monitor. Next I added the LCD display driver and sent the results to the 1602 LCD as well. It is really nice, even surprising, when everything works just as it is supposed to work!

This picture on the following page shows the resulting LCD display with the ambient and object temperatures in farenhieght.

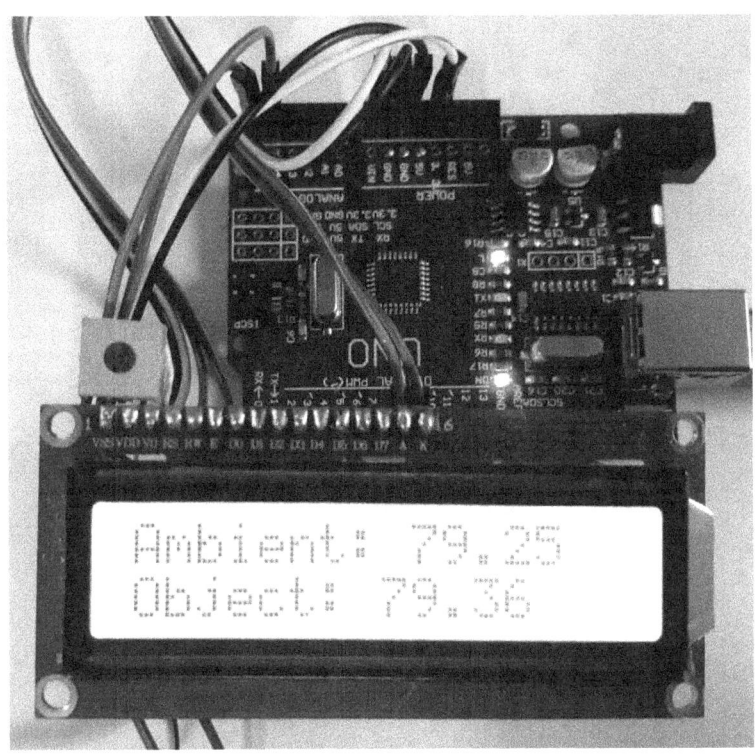

This is my code to get the results to display on a 1602 LCD screen:

```
/*****************************************************
  This is for the MLX90614 Temp Sensor
  it displays the results on a 1602 LCD
  These sensors use I2C to communicate

  Based on code by Limor Fried/Ladyada for Adafruit Industries.
  1602 LCD support added by Bob Davis
 *****************************************************/
#include <Wire.h>
#include <Adafruit_MLX90614.h>
#include <LiquidCrystal.h>
LiquidCrystal lcd(8, 9, 4, 5, 6, 7);

Adafruit_MLX90614 mlx = Adafruit_MLX90614();

void setup() {
  Serial.begin(9600);
  lcd.begin(16, 2);
```

```
  Serial.println("Adafruit MLX90614 test");
  mlx.begin();
}

void loop() {
  Serial.print("Ambient = "); Serial.print(mlx.readAmbientTempF());
  Serial.print("*F\tObject = "); Serial.print(mlx.readObjectTempF());
Serial.println("*F");
  Serial.println();

  lcd.setCursor(0, 0);
  lcd.print("Ambient: ");
  lcd.print(mlx.readAmbientTempF());
  lcd.setCursor(0, 1);
  lcd.print("Object: ");
  lcd.print(mlx.readObjectTempF());
  delay(500);
}
```

Chapter 11

AMG8833 8x8 Thermal Camera

And ILI9341 Display

This project uses an AMG8838 thermal image sensor with an ESP8266 "D1" processor board and an ILI9341 240 by 320 pixels Color LCD Display to make a basic thermal camera.

This project shows how to make a fairly simple, inexpensive thermal camera. Adafruit has lots of information on the AMG8833 8x8 thermal camera on their website. Adafruit also provides some example software for use with the AMG8833 and two examples that support using an ILI9341 LCD screen.

The Adafruit thermal camera project on their website uses a slightly different 1.4" ST7735R based LCD screen that is only 128 by 128 pixels. They also used a "Metro" processor board with their AMG8833. However their thermal camera example software supports several different processors including the ESP8266 or "D1" processor and an ILI9341 based LCD!

Our project uses the following three components and lots of jumper wires. AMG8833 -> ESP8266 (D1) -> ILI9341 LCD = Thermal Camera!

Here are the Pin connections for this ILI9341 LCD for use with a three volt D1 processor. Level shifting is not needed with the D1 processor board as it runs on 3.3 volts. One thousand ohm resistors should be used as level shifters when using a five volt processor.

LCD pin   D1 Pin
1. VCC   Connect to 3.3 volts
2. Ground   Connect to ground
3. CS   Connect to D10
4. Reset   Connect to 3.3 volts
5. D/C   Connect to D9
6. MOSI/SDI   Connect to D7
7. SCK   Connect to D5
8. LED   Connect to 3.3 Volts

The next illustration is the schematic diagram showing how to connect the LCD screen to the Arduino D1 board.

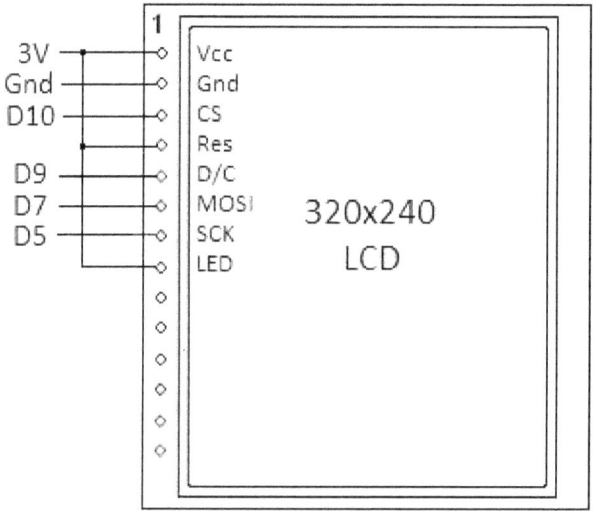

The Adafruit example programs include an interpolated pixels version. Although the camera is only 8x8 pixels in resolution this example version calculates the colors that would be between the pixels to yield many more artificial "pixels" of resolution.

The Adafruit AMG8833 tutorial is located here on the web:
https://learn.adafruit.com/adafruit-amg8833-8x8-thermal-camera-sensor
The Adafruit AMG88 drivers and example files are found here:
https://github.com/adafruit/Adafruit_AMG88xx

I made this small change, listed below, to the "interpolate" example code to get it to work with my D1 processor and with the ILI9341 LCD screen configuration. You may have to change these pin assignments to work with different LCD screens.

```
#ifdef ESP8266
   #define STMPE_CS 16
   #define TFT_CS   D10
   #define TFT_DC   D9
   #define SD_CS    2
#endif
```

The next picture is an infrared photograph of the LCD screen showing two fingers touching each other at the finger tips.

On the back side of the LCD I soldered some 30 gauge jumper wires to the power, reset and LED pins connecting the three of them together. This was done to reduce the number of jumper wires to only six that are needed to connect the LCD to the processor.

This next picture shows the AMG8833 Thermal camera. IT only takes four jumper wires to connect the thermal sensor to the processor.

This is how to connect the AMG8833 to the D1 processor:
VIN – 3.3 or 5 volts
GND – Ground
SCL – D3 (on D1 board)
SDA – D4 (on D1 board)

The next picture shows how to connect the LCD and AMG8833 to the D1 processor board. In the picture, the left two wires are from the thermal imaging camera the right four wires go to the LCD.

The LCD is powered by 3.3 volts and the thermal sensor is powered by five volts, only because there is no other 3.3 volt pins available.

I have modified the software to add a text display of the maximum temperature that has been detected. This should work well for detecting someone with a fever as they pass under the thermal camera. Basically you create two variables, scan through the readings and pick the highest temperature, then convert it to Fahrenheit and display it.

Here are the changes that are needed to the demo code to find and display the peak temperature in the upper right corner:

```
int HighTemp = 0;
int HTemp = 0;
void loop() {
 //read all the pixels
 amg.readPixels(pixels);
 Serial.print("[");
 HighTemp=0;
 for(int i=1; i<=AMG88xx_PIXEL_ARRAY_SIZE; i++){
  Serial.print(pixels[i-1]);
  Serial.print(", ");
  if( i%8 == 0 ) Serial.println();
  if (pixels[i-1] > HighTemp) HighTemp = pixels[i-1];
 }
 Serial.println("]");
 Serial.println();
 HTemp = ((HighTemp * 9/5) + 32);
 Serial.println (HTemp);

 float dest_2d[INTERPOLATED_ROWS * INTERPOLATED_COLS];
 int32_t t = millis();
 interpolate_image(pixels, AMG_ROWS, AMG_COLS, dest_2d, INTERPOLATED_ROWS, INTERPOLATED_COLS);
 Serial.print("Interpolation took "); Serial.print(millis()-t); Serial.println(" ms");

 uint16_t boxsize = min(tft.width() / INTERPOLATED_COLS, tft.height() / INTERPOLATED_COLS);
```

```
  drawpixels(dest_2d, INTERPOLATED_ROWS,
INTERPOLATED_COLS, boxsize, boxsize, false);
}

void drawpixels(float *p, uint8_t rows, uint8_t cols, uint8_t boxWidth,
uint8_t boxHeight, boolean showVal) {
  int colorTemp;
  for (int y=0; y<rows; y++) {
    for (int x=0; x<cols; x++) {
      float val = get_point(p, rows, cols, x, y);
      if(val >= MAXTEMP) colorTemp = MAXTEMP;
      else if(val <= MINTEMP) colorTemp = MINTEMP;
      else colorTemp = val;
uint8_t colorIndex = map(colorTemp, MINTEMP, MAXTEMP, 0, 255);
      colorIndex = constrain(colorIndex, 0, 255);
      //draw the pixels!
      uint16_t color;
      color = val * 2;
      tft.fillRect(boxWidth * x, boxHeight * y, boxWidth, boxHeight,
camColors[colorIndex]);

      if (showVal) {
        tft.setCursor(boxWidth * y + boxWidth/2 - 12, 40 + boxHeight * x +
boxHeight/2 - 4);
        tft.setTextColor(ILI9341_WHITE);  tft.setTextSize(2);
        tft.print(val,1);
      }
    }
  }
  tft.setTextSize(2);
  tft.setTextColor(ILI9341_WHITE, ILI9341_BLACK);
  tft.setCursor(rows*boxWidth,0);
  tft.print(" High");
  tft.setCursor(rows*boxWidth,20);
  tft.print(" Temp:");
  tft.setCursor(rows*(boxWidth+1),40);
  tft.print(HTemp);
  tft.print(" ");
}
```

Chapter 12

Temperature and Humidity

To OLED Display

In this chapter we will first introduce the 96 x 64 Full Color OLED display. Once again I was able to get the pins to line up so that a header extender can be used to connect the Arduino to the OLED display. The pin alignment requires that the OLED overlap the Arduino, but the use of a header makes it much easier to add the display. One pin needs to be bent on the five pin header to skip over the Arduino D12 pin. Then I used short jumper wires for the OLED power and ground as seen in the diagram below.

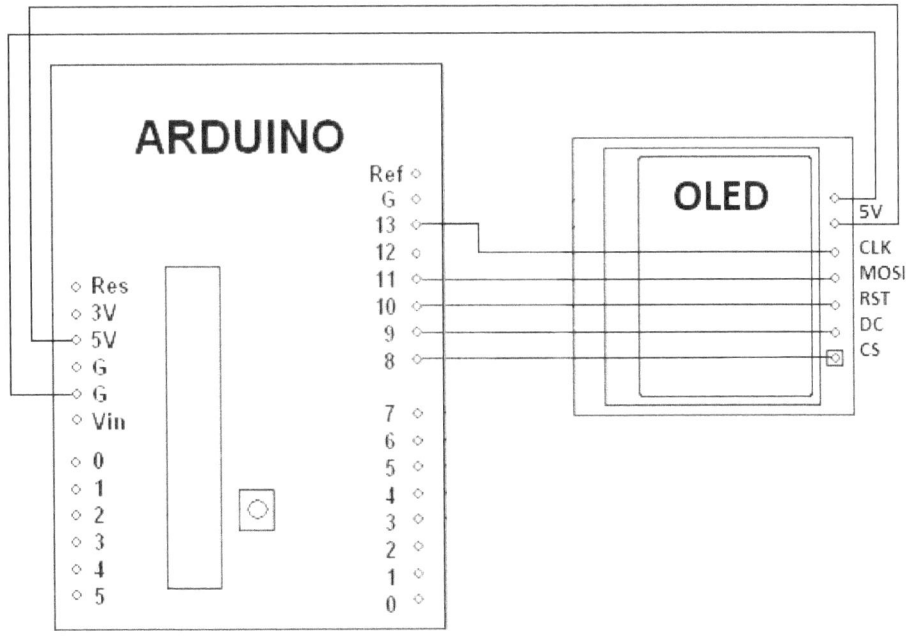

OLED Displays look just like a LCD screen, but each pixel is actually an Organic LED. Because the pixels are LED's you do not need a backlight like LCD displays use. The use of LED's also makes the screen much brighter and the colors are stronger than with a LCD screen.

You will need to add the SSD1331 library using the library manager or download it and copy the unzipped file to your Arduino/Library folder.

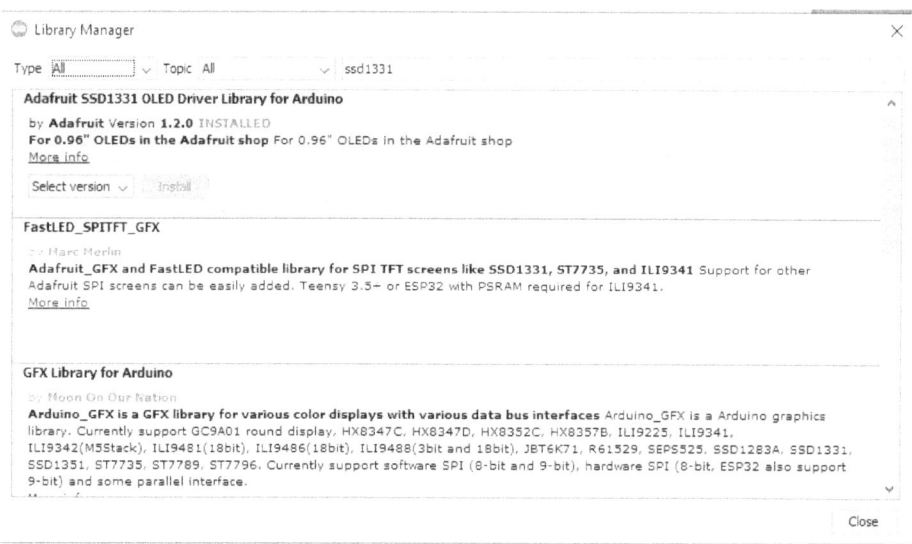

Next you will need two DHT11's for this project. The DHT 11 comes in a three pin or four pin case and can also be purchased mounted on a small circuit board. The left pin is power the next is the signal and the right most pin is ground. I soldered jumper wires on my DHT11's to make easier to connect them up to the Arduino. The signal wire from one DHT11 goes to the Arduino D2 and the other DHT11 signal wire goes to D3

The next picture is of a DHT11 temperature and humidity sensor with the three wires attached to it and covered in heat shrink insulators.

Next you will need to install the DHT Sensor Library. You can find the DHT11 support software in the Arduino Library manager.

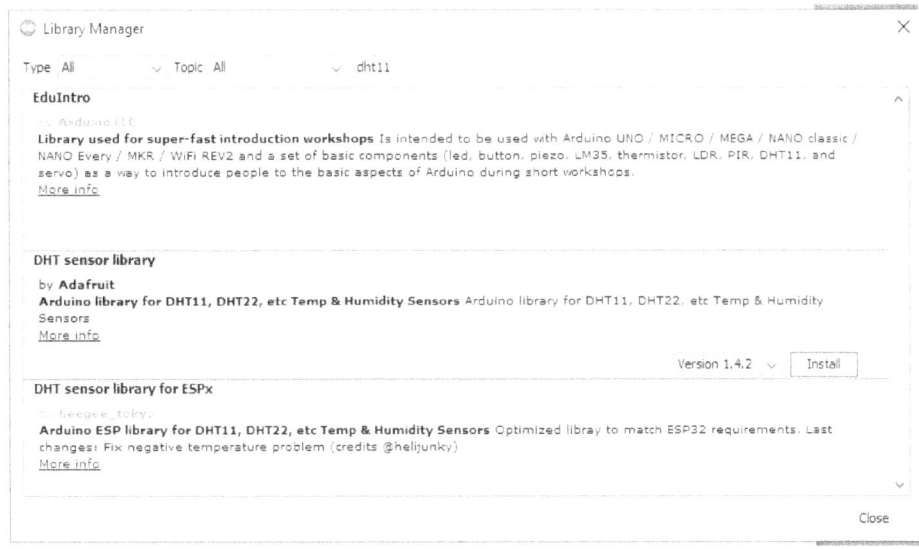

Then you can download my code from github or type it in from the code below. If everything is working you should see the Indoor and Outdoor temperatures in Fahrenheit and humidity percent being displayed on the OLED display.

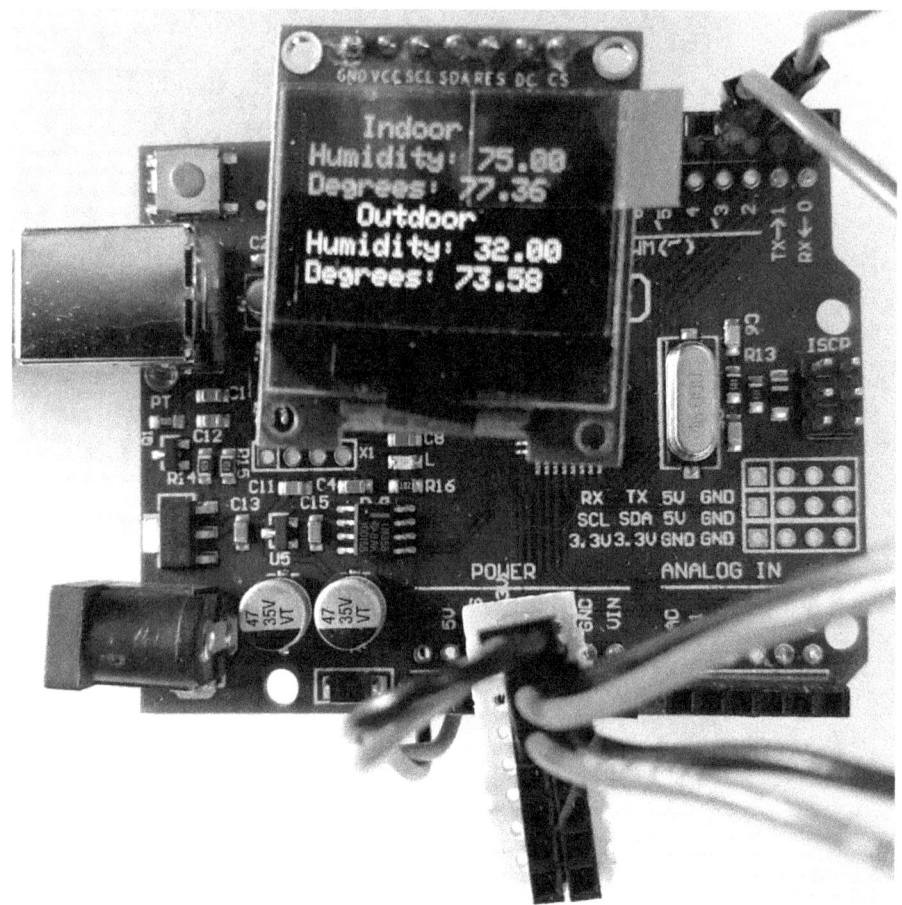

This is the code for a basic OLED display, there is a more complex graphing version at my github account under bobdavis321.

// 2 DHT11 to OLED

// Example testing sketch for various DHT humidity/temperature sensors
// Written by ladyada, public domain
// Modified for OLED and dual sensors by Bob Davis on 4-26-2021
// REQUIRES the following Arduino libraries:
// - DHT Sensor Library: https://github.com/adafruit/DHT-sensor-library

// - Adafruit Unified Sensor Lib: https://github.com/adafruit/Adafruit_Sensor

// Connect pin 1 (on the left) of the sensor to +5V
// Connect pin 2 of the sensor to whatever your DHTPIN 2 and 3
// Connect pin 3 (on the right) of the sensor to GROUND (if your sensor has 3 pins)
// Connect pin 4 (on the right) of the sensor to GROUND and leave the pin 3 EMPTY (if your sensor has 4 pins)

#include "DHT.h"
#include <Adafruit_GFX.h>
#include <Adafruit_SSD1331.h>
#include <SPI.h>

// You can use any (4 or) 5 pins
// This pin arrangement makes possible the use of a header extender
#define sclk 13
#define mosi 11
#define rst 10
#define dc   9
#define cs   8

// Color definitions
#define BLACK   0x0000
#define BLUE   0x001F
#define RED   0xF800
#define GREEN   0x07E0
#define CYAN   0x07FF
#define MAGENTA   0xF81F
#define YELLOW   0xFFE0
#define WHITE   0xFFFF

#define DHTPIN 2   // Digital pin connected to the DHT sensor
#define DHTPIN2 3   // Digital pin connected to the DHT2 sensor

// Un-comment whatever type of sensor you're using!
#define DHTTYPE DHT11   // DHT 11
//#define DHTTYPE DHT22   // DHT 22 (AM2302), AM2321
//#define DHTTYPE DHT21   // DHT 21 (AM2301)

```
// Initialize DHT sensors, OLED.
DHT dht(DHTPIN, DHTTYPE);
DHT dht2(DHTPIN2, DHTTYPE);
Adafruit_SSD1331 display = Adafruit_SSD1331(&SPI, cs, dc, rst);

void setup() {
  Serial.begin(9600);
  Serial.println(F("DHT test!"));
  display.begin();
  display.fillScreen(BLACK);
  dht.begin();
  dht2.begin();
}

void loop() {
  // Wait a few seconds between measurements.
  delay(2000);

  // Reading temperature or humidity takes about 250 milliseconds!
  // Sensor readings may also be up to 2 seconds 'old' (its a very slow sensor)
  float hum = dht.readHumidity();
  float hum2 = dht2.readHumidity();
  // Read temperature as Fahrenheit (isFahrenheit = true)
  float tem = dht.readTemperature(true);
  float tem2 = dht2.readTemperature(true);

  // Check if read failed and try again.
  if (isnan(hum) || isnan(tem)) {
    Serial.println(F("Failed to read from DHT sensor!"));
    return;
  }
  display.fillScreen(BLACK);
  display.setTextColor(RED);
  display.setCursor(0,0);
  display.print("   Indoor ");
  display.setCursor(0,10);
  display.print("Humidity: ");
  display.print(hum);
  display.setCursor(0,20);
  display.print("Degrees: ");
```

```
  display.print(tem);

  display.setTextColor(GREEN);
  display.setCursor(0,30);
  display.print("   Outdoor ");
  display.setCursor(0,40);
  display.print("Humidity: ");
  display.print(hum2);
  display.setCursor(0,50);
  display.print("Degrees: ");
  display.print(tem2);

  // For serial monitor:
  Serial.print(F("Humidity: "));
  Serial.print(hum);
  Serial.print(F("%  Temperature: "));
  Serial.print(tem);
  Serial.print(F("°F"));
  Serial.print('\n');  //new line
}
```

This program can also work with a black and white OLED display as can be seen in the next picture.

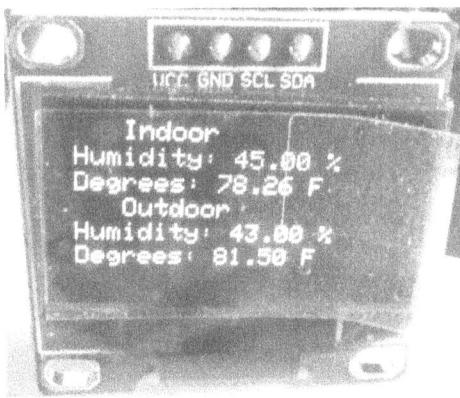

The SSD1306 easily connects to SCL, SDA, 3.3 volts and ground. Only four wires are needed to connect it to an Arduino.

Here is the code for a SSD1306

/***

```
// SSD1306 displaying two DHT11's

// Example testing sketch for various DHT humidity/temperature sensors
// Written by ladyada, public domain
// Modified for OLED and dual sensors by Bob Davis on 4-26-2021
// REQUIRES the following Arduino libraries:
// - DHT Sensor Library: https://github.com/adafruit/DHT-sensor-library
// - Adafruit Unified Sensor Lib:
https://github.com/adafruit/Adafruit_Sensor
// Connect pin 1 (on the left) of the sensor to +5V
// Connect pin 2 of the sensor to whatever your DHTPIN 2 and 3
// Connect pin 3 (on the right) of the sensor to GROUND if 3 pin sensor
// Connect pin 4 (on the right) of the sensor to GROUND
************************************************************/
#include "DHT.h"
#include <SPI.h>
#include <Wire.h>
#include <Adafruit_GFX.h>
#include <Adafruit_SSD1306.h>

#define SCREEN_WIDTH 128 // OLED display width, in pixels
#define SCREEN_HEIGHT 64 // OLED display height, in pixels

#define DHTPIN 2    // Digital pin connected to the DHT sensor
#define DHTPIN2 3   // Digital pin connected to the DHT2 sensor

// Uncomment whatever type of sensor you're using!
#define DHTTYPE DHT11    // DHT 11
//#define DHTTYPE DHT22   // DHT 22  (AM2302), AM2321
//#define DHTTYPE DHT21   // DHT 21 (AM2301)

// Initialize DHT sensors.
DHT dht(DHTPIN, DHTTYPE);
DHT dht2(DHTPIN2, DHTTYPE);

// Declaration for an SSD1306 display connected to I2C (SDA, SCL pins)
// The pins for I2C are defined by the Wire-library.
// On an arduino UNO:       A4(SDA), A5(SCL)
// On an arduino MEGA 2560: 20(SDA), 21(SCL)
// On an arduino LEONARDO:   2(SDA),  3(SCL), ...
```

```cpp
#define OLED_RESET     4 // Reset pin # (or -1 if sharing Arduino reset pin)
#define SCREEN_ADDRESS 0x3C // See datasheet for Address; 0x3D=128x64, 0x3C=128x32
Adafruit_SSD1306 display(SCREEN_WIDTH, SCREEN_HEIGHT, &Wire, OLED_RESET);

void setup() {
  Serial.begin(9600);
  Serial.println(F("DHT test!"));
  display.begin();
  display.fillScreen(BLACK);
  dht.begin();
  dht2.begin();

  // SSD1306_SWITCHCAPVCC = generate display voltage from 3.3V internally
  if(!display.begin(SSD1306_SWITCHCAPVCC, SCREEN_ADDRESS)) {
    Serial.println(F("SSD1306 allocation failed"));
    for(;;); // Don't proceed, loop forever
  }

  // Clear the buffer
  display.clearDisplay();
}

void loop() {
  // Wait a few seconds between measurements.
  delay(2000);

  // Reading temperature or humidity takes about 250 milliseconds!
  float hum = dht.readHumidity();
  float hum2 = dht2.readHumidity();
  // Read temperature as Fahrenheit (isFahrenheit = true)
  float tem = dht.readTemperature(true);
  float tem2 = dht2.readTemperature(true);

  // Check if read failed and try again.
  if (isnan(hum) || isnan(tem)) {
    Serial.println(F("Failed to read from DHT sensor!"));
```

```
  return;
}
display.clearDisplay();
display.setTextColor(SSD1306_WHITE);
display.setCursor(0,0);
display.print("   Indoor ");
display.setCursor(0,10);
display.print("Humidity: ");
display.print(hum);
display.print(" % ");
display.setCursor(0,20);
display.print("Degrees: ");
display.print(tem);
display.print(" F ");

display.setTextColor(SSD1306_WHITE);
display.setCursor(0,30);
display.print("   Outdoor ");
display.setCursor(0,40);
display.print("Humidity: ");
display.print(hum2);
display.print(" % ");
display.setCursor(0,50);
display.print("Degrees: ");
display.print(tem2);
display.print(" F ");

display.display();
// For serial monitor:
Serial.print("Humidity: ");
Serial.print(hum);
Serial.print(" ");
Serial.print(hum2);
Serial.print("%  Temperature: ");
Serial.print(tem);
Serial.print(" ");
Serial.print(tem2);
Serial.print("°F");
Serial.print('\n');  //new line

}
```

Chapter 13

Temperature and Humidity

to a Webpage

Now, let us take what we have made so far to another level. We are going to add network connectivity so that you can see the temperature and humidity on your telephone or computer! To do this we will use an ESP8266 or D1 mini processor. It is roughly the same as the D1 but it is in a much smaller package with less I/O pins that are available.

First I added the SSD1306 OLED display to the D1 mini. This turned out to be a real headache to get it working! For some reason you cannot pass the normal parameters from a D1 to the SSD1306! As a result we lose every other line on the screen.

After many hours of trying to fix the issue, including modifying the support library, and other desperate attempts, I gave up. I settled on using the screen in that limited, every other line arrangement. So there are only 32 lines instead of 64 lines on the display. This required modifying the text to fit both the indoor and outdoor readings into 32 lines of resolution.

The SSD1306 display's SCL and SDA are connected to the pins marked D1 and D2 on the D1 mini. To run two DHT11's you will need two sets of 5 volt and ground pins. I used my own home made adapter for that purpose. This power adapter can be seen on the upper right corner in the next picture.

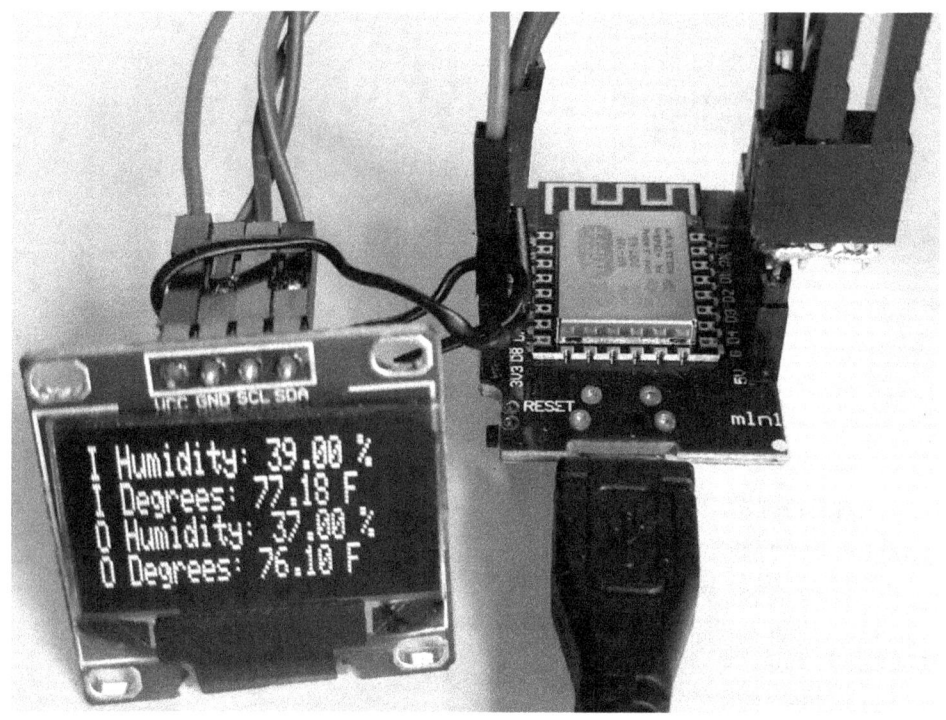

When you run the code, you will need to watch the Arduino serial monitor to see what IP address has been assigned to the D1 processor. Then put that IP address into your web browser and you can then see the indoor and outdoor temperature and humidity readings as seen below. The font is a little on the small size on my telephone.

```
ESP8266 D1 Indoor+Outdoor

Indoor

Temperature 76.28

Humidity 40.00

Outdoor

Temperature 76.46

Humidity 38.00
```

I could have added code to make the display a little fancier but to limit the size of this book, the software is limited in size. Here is the code to make the D1 work with the OLED and as a server on a Network.

```
/***********************************************************
D1 SSD1306 displaying two DHT11's and sending results to webpage

// Example testing sketch for various DHT humidity/temperature sensors
// Written by ladyada, public domain
// Modified for OLED and dual sensors by Bob davis on 4-26-2021
// REQUIRES the following Arduino libraries:
// - DHT Sensor Library: https://github.com/adafruit/DHT-sensor-library
// - Adafruit Unified Sensor Lib: https://github.com/adafruit/Adafruit_Sensor

// Connect pin 1 (on the left) of the sensor to +5V
// Connect pin 2 of the sensor to whatever your DHTPIN 2 and 3
// Connect pin 3 (on the right) of the sensor to GROUND (if your sensor has 3 pins)
// Connect pin 4 (on the right) of the sensor to GROUND and leave the pin 3 EMPTY

***********************************************************/
#include "DHT.h"
#include <SPI.h>
#include <Wire.h>
#include <Adafruit_GFX.h>
#include <Adafruit_SSD1306.h>
#include <ESP8266WiFi.h>
#include <WiFiClient.h>
#include <ESP8266WebServer.h>

// Set WiFi credentials - use your credentials
#define WIFI_SSID "BobDavis"
#define WIFI_PASS "XXXXXXX"

#define DHTPIN D5     // Digital pin connected to the DHT sensor
#define DHTPIN2 D6    // Digital pin connected to the DHT2 sensor

// Uncomment whatever type of sensor you're using!
#define DHTTYPE DHT11   // DHT 11
```

```cpp
//#define DHTTYPE DHT22   // DHT 22  (AM2302), AM2321
//#define DHTTYPE DHT21   // DHT 21 (AM2301)

float tem1;
float tem2;
float hum1;
float hum2;

// Initialize DHT sensors
DHT dht(DHTPIN, DHTTYPE);
DHT dht2(DHTPIN2, DHTTYPE);
// Initialize webserver
ESP8266WebServer webserver(80);

// Declaration for an SSD1306 display connected to I2C (SDA, SCL pins)
// The pins for I2C are defined by the Wire-library.
// On an arduino UNO:       A4(SDA), A5(SCL)
// On an arduino MEGA 2560: 20(SDA), 21(SCL)
// On an arduino LEONARDO:   2(SDA),  3(SCL)
// On an arduino D1:        D2(SDA), D1(SCL)
#define OLED_RESET    0 // Reset pin # (or -1 if sharing Arduino reset pin)
#define SCREEN_ADDRESS 0x3C // See datasheet for Address; 0x3D=128x64, 0x3C=128x32
Adafruit_SSD1306 display(OLED_RESET); // Works with D1 for some reason
//Adafruit_SSD1306 display(SCREEN_WIDTH, SCREEN_HEIGHT, &Wire, OLED_RESET);

void setup() {
  Serial.begin(9600);
  Serial.println(F("DHT test!"));
  display.begin();
  display.clearDisplay();
  display.setTextSize(1);
  display.setTextColor(WHITE);
  dht.begin();
  dht2.begin();

  WiFi.begin(WIFI_SSID, WIFI_PASS);
  while (WiFi.status() != WL_CONNECTED) { delay(100); }
```

```
  // WiFi Connected
  Serial.print("Connected! IP address: ");
  Serial.println(WiFi.localIP());

  // Start Web Server
  webserver.on("/", rootPage);
  webserver.begin();
}

void loop() {
  webserver.handleClient();
  // Wait a few seconds between measurements.
  delay(2000);

  // Reading temperature or humidity takes about 250 milliseconds!
  // Sensor readings may also be up to 2 seconds 'old' (its a very slow sensor)
  hum1 = dht.readHumidity();
  hum2 = dht2.readHumidity();
  // Read temperature as Fahrenheit (isFahrenheit = true)
  tem1 = dht.readTemperature(true);
  tem2 = dht2.readTemperature(true);

  display.clearDisplay();
  display.setCursor(0,0);
  display.print("I Humidity: ");
  display.print(hum1);
  display.print(" % ");
  display.setCursor(0,8);
  display.print("I Degrees: ");
  display.print(tem1);
  display.print(" F ");

  display.setCursor(0,16);
  display.print("O Humidity: ");
  display.print(hum2);
  display.print(" % ");
  display.setCursor(0,24);
  display.print("O Degrees: ");
  display.print(tem2);
```

```
  display.print(" F ");

  display.display();
  // For serial monitor:
  Serial.print("Humidity: ");
  Serial.print(hum1);
  Serial.print(" ");
  Serial.print(hum2);
  Serial.print("%  Temperature: ");
  Serial.print(tem1);
  Serial.print(" ");
  Serial.print(tem2);
  Serial.print("°F");
  Serial.print('\n');  //new line
}

void rootPage() {
  String strTemp, strHum, strTemp2, strHum2;
  char tmp[10];

  dtostrf(tem1,1,2,tmp);
  strTemp = tmp;
  dtostrf(hum1,1,2,tmp);
  strHum = tmp;
  dtostrf(tem2,1,2,tmp);
  strTemp2 = tmp;
  dtostrf(hum2,1,2,tmp);
  strHum2 = tmp;

  webserver.send(200,"text/plain","ESP8266 D1 Indoor+Outdoor\n\rIndoor\n\rTemperature "+strTemp+"\n\rHumidity "+strHum+ "\n\rOutdoor\n\rTemperature "+strTemp2+"\n\rHumidity "+strHum2);
}
```

Chapter 14

Temperature, Pressure

and Humidity to OLED

For this chapter we are going to introduce the BME280. To avoid confusion there is also a BMP280 that does not give humidity. The two are often confused with each other. I wrote a program for the BMP280 that is available on my github account under bobdavis321. The two sensors are NOT software interchangeable.

The BME280 uses the I2C interface also known as A4 and A5 or also known as SDA and SCL. It also needs 3.3 volts and ground to operate. I chose the SSD1331 OLED display for its bright colors. You could easily change the software for a black and white OLED display.

The BME280 uses SCL and SCA but the SSD1331 uses five pins for a software based driver. It needs a Clock (sclk), Serial data (mosi), reset (rst), data/control (dc) selection and chip selection (cs).

This is the code for the BME280 to color SSD1331 OLED configuration.

```
// BME-280 to OLED
// 6-22-2021 by Bob Davis
// Uses Three Adafruit Libraries

#include <Adafruit_GFX.h>
#include <Adafruit_SSD1331.h>
#include <Adafruit_BME280.h>
#include <SPI.h>
#include <Wire.h>

// You can use any five pins
// This pin arrangement makes possible the use of a header extender
#define sclk 13
#define mosi 11
#define rst  10
#define dc   9
#define cs   8

// Color definitions
#define BLACK      0x0000
#define BLUE       0x001F
#define RED        0xF800
#define GREEN      0x07E0
#define CYAN       0x07FF
#define MAGENTA    0xF81F
#define YELLOW     0xFFE0
#define WHITE      0xFFFF

// Initialize Sensors, OLED.
Adafruit_SSD1331 display = Adafruit_SSD1331(&SPI, cs, dc, rst);
Adafruit_BME280 bme; // use I2C interface
Adafruit_Sensor *bme_temp = bme.getTemperatureSensor();
Adafruit_Sensor *bme_pressure = bme.getPressureSensor();
Adafruit_Sensor *bme_humidity = bme.getHumiditySensor();
```

```
void setup() {
  Serial.begin(9600);
  Serial.println(F("BME280 test!"));
  display.begin();
  display.fillScreen(BLACK);
  if (!bme.begin()) {
    Serial.println(F("Could not find a valid BME280 sensor"));
  }
  bme_temp->printSensorDetails();
  bme_pressure->printSensorDetails();
  bme_humidity->printSensorDetails();
}

void loop() {
  sensors_event_t temp_event, pressure_event, humidity_event;
  bme_temp->getEvent(&temp_event);
  bme_pressure->getEvent(&pressure_event);
  bme_humidity->getEvent(&humidity_event);

  Serial.print(F("Temperature = "));
  Serial.print(temp_event.temperature);
  Serial.println(" *C");
  Serial.print(F("Humidity = "));
  Serial.print(humidity_event.relative_humidity);
  Serial.println(" %");
  Serial.print(F("Pressure = "));
  Serial.print(pressure_event.pressure);
  Serial.println(" hPa");
  Serial.println();

  display.fillScreen(BLACK);
  display.setTextColor(RED);
  display.setCursor(0,0);
  display.print("Temp. C: ");
  display.print(temp_event.temperature);
  display.setTextColor(BLUE);
  display.setCursor(0,15);
  display.print("Temp. F: ");
  display.print((temp_event.temperature*9/5)+32);
  display.setTextColor(GREEN);
  display.setCursor(0,30);
```

```
display.print("Pressure: ");
display.print(pressure_event.pressure);
display.setTextColor(YELLOW);
display.setCursor(0,45);
display.print("Humidity: ");
display.print(humidity_event.relative_humidity);
delay(2000);
}
```

Chapter 15

Biometric Watch

When I started this book I had the idea of making a watch that would monitor your heart, activity, etc. However, I soon found out that someone already makes such a device and it sells for about $30. You cannot make one for that price that I know of. However, it is a lot of fun and very educational to make one of these for yourself. So, for this project we are going to put all that we have learned in the previous chapters together in one project, and then add some new technology as well.

We will need to add a real time clock. You can do that with software as well but the Dallas DS1307 is just one tiny chip that will take care of the time, date and year as well. It even figures out the correct number of days in each month according to the year. The DS1307 can have a backup battery to maintain the time without the main battery.

I wanted to add an accelerometer for fitness and activity tracking but was a little short on room. It could be added later.

The computer for this project is a D1 mini. I was hoping that a lithium battery, that I had on hand, would fit between the D1 mini's rows of pins, but it did not fit.

This whole project would be better built on a printed circuit board. I used a fiberglass prototype board with holes on .1 inches centers. I had to carve a notch into it for the SSD1307 OLED display so it could be mounted almost flush. Below that there are three switches for "Up", "Enter", and "Down". On the back of the board the DS1307 is glued in place, the D1 mini and the MAX30102 are socketed.

All the devices in this project are 3.3 volt friendly and use SCL and SDA. To a degree this simplifies things. The needed pull-up resistors are on the MAX30102 heartbeat detector.

This is the schematic diagram of the watch.

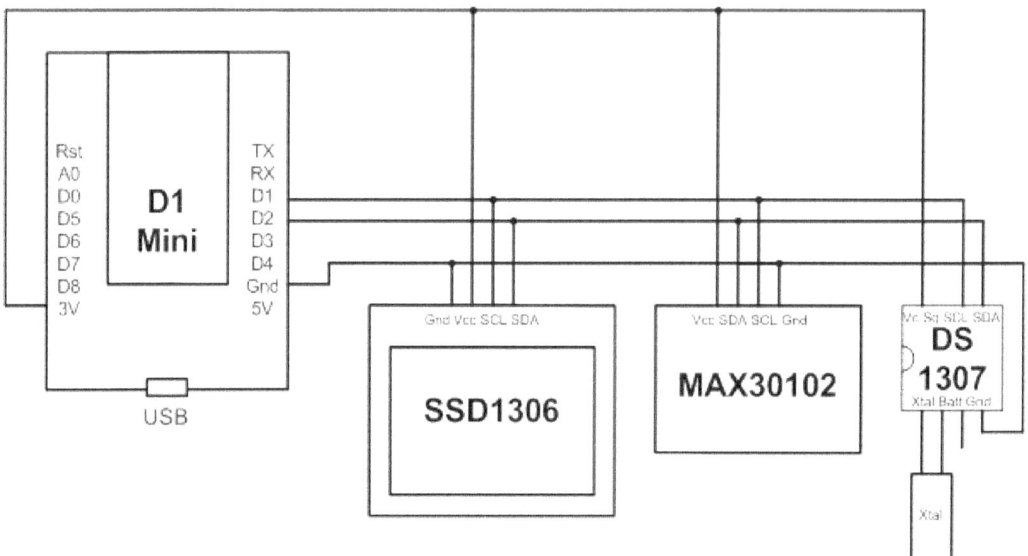

Here is a picture of the watch displaying day of the week, the date and the time.

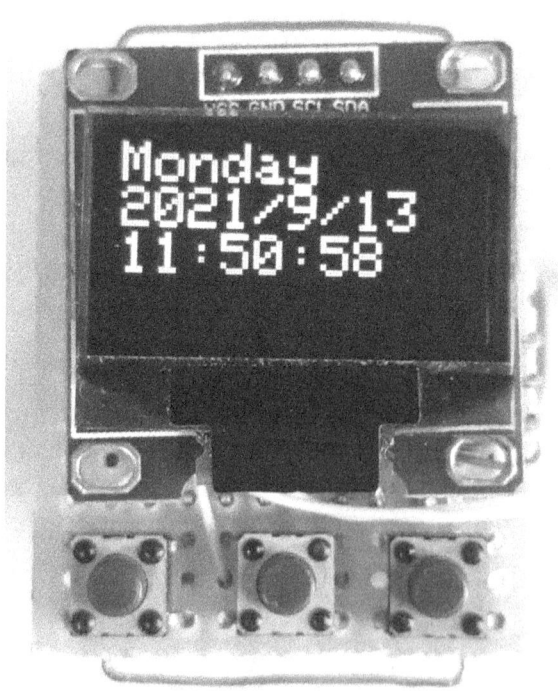

The next two pictures show the circuit board top and bottom views. The display connector is cut into the board then it is glued in place.

The driver that I used for the DS1307 is called "RTClib" by Adafruit. The code is based on the demo program that comes with the driver.

RTClib
by Adafruit version 1.14.1 INSTALLED
A fork of Jeelab's fantastic RTC library Works with DS1307, DS3231, PCF8523, PCF8563 on multiple architectures
More info

The code for the watch is still under development. Here is the code for the time keeping portion.

```
// Arduino D1 Mini with SSD1306 and Dallas RTC DS1307 watch.
// By Bob Davis on 9/14/2021
// Based on code from Adafruit
//
// Plan to add MAX1302 Heartbeat detection and accelerometer.
//
#include <SPI.h>
#include <Wire.h>
#include <Adafruit_GFX.h>
#include <Adafruit_SSD1306.h>
#include "RTClib.h"
RTC_DS1307 rtc;
char daysOfTheWeek[7][12] = {"Sunday", "Monday", "Tuesday", "Wednesday", "Thursday", "Friday", "Saturday"};

#define SCREEN_WIDTH 128 // OLED display width, in pixels
#define SCREEN_HEIGHT 64 // OLED display height, in pixels
#define OLED_RESET 0  // Normally 4 => UNO
#define SCREEN_ADDRESS 0x3C
//Adafruit_SSD1306 display(OLED_RESET);
Adafruit_SSD1306 display(SCREEN_WIDTH, SCREEN_HEIGHT, &Wire, OLED_RESET);

void setup()  {
 Serial.begin(9600);
 // SSD1306_SWITCHCAPVCC = generate display voltage from 3.3V internally
 if(!display.begin(SSD1306_SWITCHCAPVCC, SCREEN_ADDRESS))
 {
```

```
    Serial.println(F("SSD1306 allocation failed"));
    for(;;); // Don't proceed, loop forever
  }
#ifndef ESP8266
  while (!Serial); // wait for serial port to connect. Needed for native USB
#endif

  if (! rtc.begin()) {
    Serial.println("Couldn't find RTC");
    Serial.flush();
    abort();
  }

  if (! rtc.isrunning()) {
    Serial.println("RTC is NOT running, let's set the time!");
    // When time needs to be set on a new device, or after a power loss, the
    // following line sets the RTC to the date & time this sketch was compiled
    rtc.adjust(DateTime(F(__DATE__), F(__TIME__)));
  }
}

void loop() {
    DateTime now = rtc.now();
    Serial.print(now.year(), DEC);
    Serial.print('/');
    Serial.print(now.month(), DEC);
    Serial.print('/');
    Serial.print(now.day(), DEC);
    Serial.print(" (");
    Serial.print(daysOfTheWeek[now.dayOfTheWeek()]);
    Serial.print(") ");
    Serial.print(now.hour(), DEC);
    Serial.print(':');
    Serial.print(now.minute(), DEC);
    Serial.print(':');
    Serial.print(now.second(), DEC);
    Serial.println();

    //Add stuff into the 'display buffer'
    display.clearDisplay();
```

```
  display.setTextSize(2);
  display.setTextColor(SSD1306_WHITE);
  display.setCursor(0,0);
  display.println(daysOfTheWeek[now.dayOfTheWeek()]);
  // Line 2
  display.print(now.year(), DEC);
  display.print('/');
  display.print(now.month(), DEC);
  display.print('/');
  display.println(now.day(), DEC);
  // Line 3
  display.print(now.hour(), DEC);
  display.print(':');
  display.print(now.minute(), DEC);
  display.print(':');
  display.print(now.second(), DEC);
  // Send to display
  display.display(); //output 'display buffer' to screen

  delay(1000);
}
```

Chapter 16

Arduino Colloidal Silver Maker

I have used a colloidal silver maker for years, but it was not mine, and the owner asked for it back. Then I looked for my home made colloidal silver maker but could not find it. I have way too much junk to look through. Then I got an idea, why not make a colloidal silver maker that is powered by an Arduino so it could control the voltage, current, reverse the polarity, record the results and even time the operation.

This is my latest schematic of the interface to the colloidal silver maker. I used a L293 motor controller because it supports reversing the polarity and can easily run the stirrer motor.

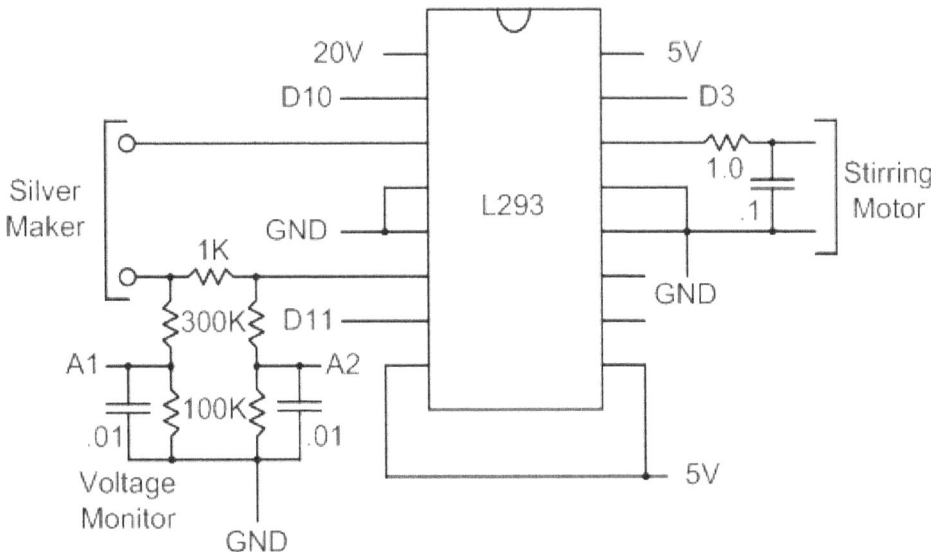

I had to add some filter capacitors in the circuit to attempt to get more stable numbers for the LCD display and to prevent premature shut off. I have changed the resistors in the divider to 300K (Two 150K in series) and 100K to further reduce the current reading when there is no water is present. Also note that the stirrer motor is now on D3 for PWM speed control ability. Eventually you might be able to adjust the stirrer motors voltage with a few key presses. I have also made a small circuit board with the voltage and current monitoring resistors on it. This circuit board is visible in some of the latest pictures.

This is the schematic diagram showing the 1602 LCD wiring. This is identical to the LCD shield wiring commonly available on eBay. This is the LCD wiring as seen from the bottom side!

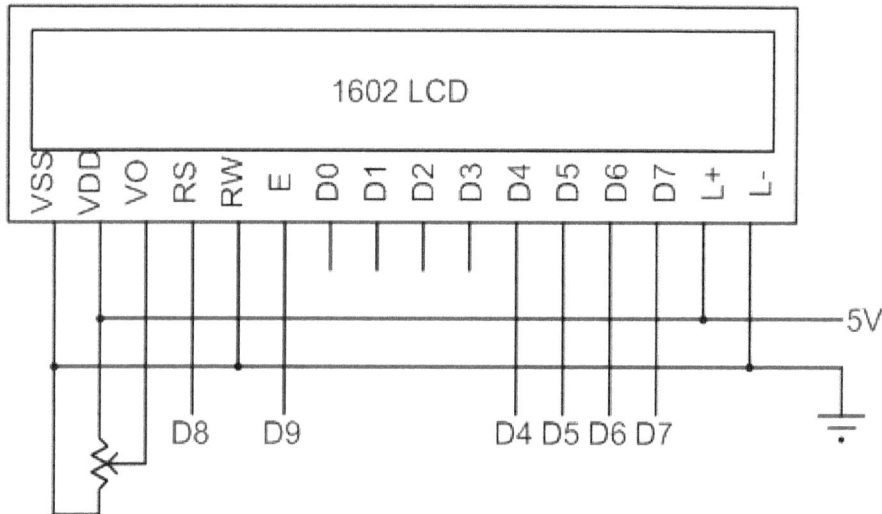

This next picture is of what the LCD screen looks like in an earlier version. I could use a bigger screen! The LCD is saying the voltage is .51 volts and .25 volts across a 1K ohm resistor for a current of .25ma; it usually never reaches one ma during over four hours of operation. Every 30 minutes the Arduino reverses the polarity and the LCD will then show around 17.7 volts with a 19 volt power source. The bottom line of the LCD also displays the running time.

This is what the LCD looks like after over 30 minutes of operation, when the polarity has been reversed by the Arduino. With a 19 volt power source you can see that between 17 and 18 volts make it to the colloidal silver maker. The voltage and current will bounce around a lot, likely because of the noise from the air pump motor.

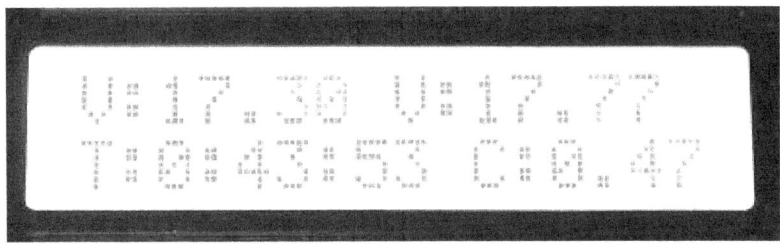

Up next is a picture of the colloidal silver maker, when it was still using a breadboard for the wiring. The L293 motor control board is visible on the right side of the picture. This picture shows the Colloidal Silver Generator while it is actually running. The Air pump stirrer has been added with a 20 ohm resistor in series (Not 1 ohm as was in the schematic diagram) to reduce the motor noise. Also the motor ground must be kept separate from the other grounds because of all the electrical noise that the motor makes.

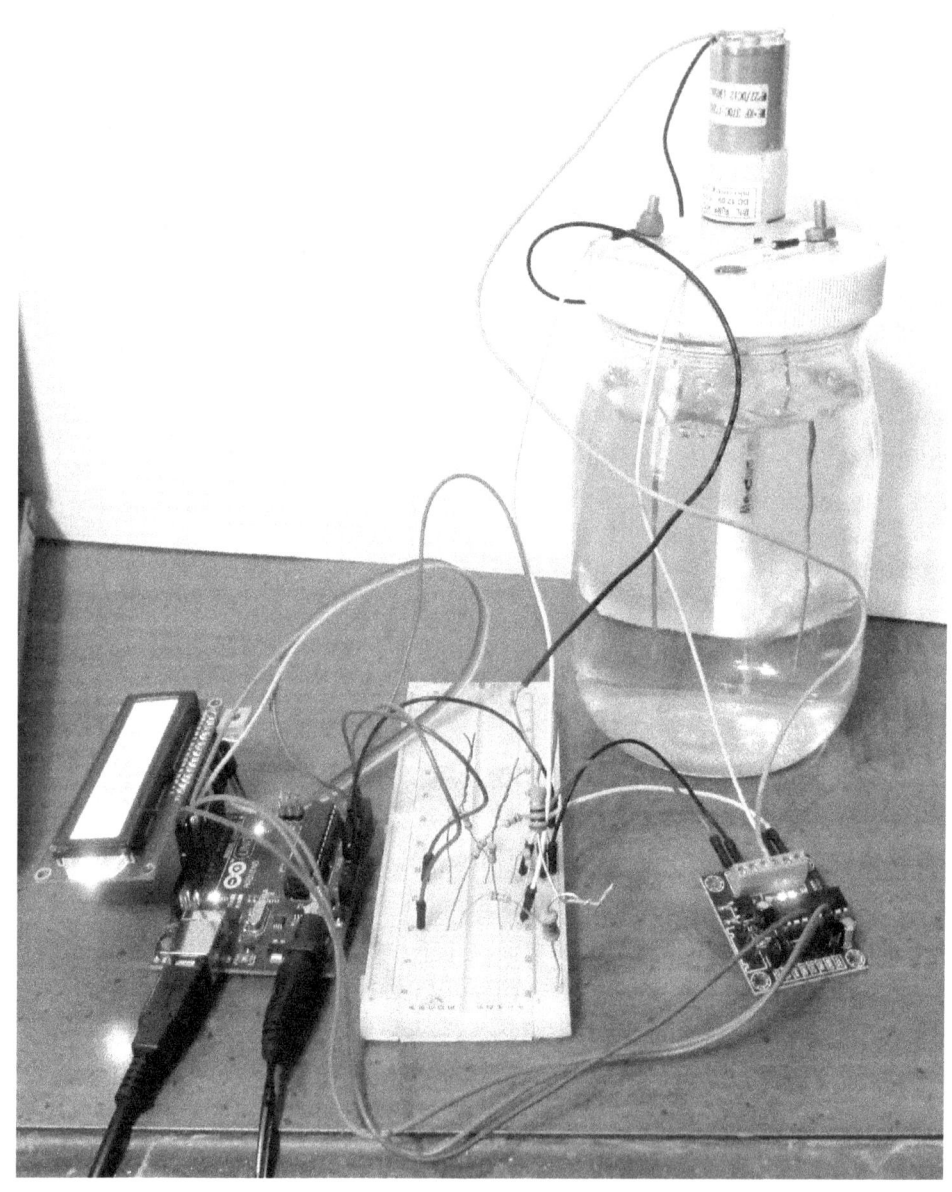

On the next page is another picture, this time it is the completed colloidal silver maker. The resistor divider and L293 motor controller are now located on the top of the colloidal silver maker. Eight wires then run down to the Arduino. There are two ground wires, one ground wire for the motor controller and one ground wire for the voltage divider.

After the second test run at 12 volts with the current staying under .5 ma, and 1 ma being the ideal current, it needed to be modified from 12 volts to 20 volt operation. Most Colloidal silver makers use 24 to 28 volts. The L293 can operate up to 30 volts, but the Arduino voltage regulator has a maximum of 20 volts, and the air pump stirrer has a maximum of 12 volts. The air pump runs best at around nine volts. So it is best to try to use a 19 to 20 volt power source like that of an old laptop computers ac adapter.

I tested this design for a few minutes with a 19 volt laptop AC adapter. Only 17.7 volts made it to the colloidal silver maker. The Arduino voltage regulator got very warm but it survived. I disconnected the air pump for this test because it is rated for 12 volts maximum and runs best at around 7-9 volts. I have since added a PWM output from the Arduino for the air pump of 1/2 of the power source or 10 volts for 20 volt operation.

The Arduino's PWM ability is used to regulate the current to the Colloidal Silver maker to just under 1 ma. The over-current shutdown is set to kick in at 2.0 ma.

Here is the code so far:

```
/****************************
Arduino Colloidal Silver Maker
By Bob Davis on April 2020

Uses a 16x2 LCD display shield or equivalent
Shows the voltage, current, and run time.

The circuit:
 * LCD RS - D9
 * LCD Enable - D8
 * LCD D4 - D4
 * LCD D5 - D5
 * LCD D6 - D6
 * LCD D7 - D7
 * LCD R/W and VSS pin to ground
 * LCD VCC and LED pin to 5V
```

* Uses L293 motor controller on D10 and D11 for PWM ability
 * Uses L293 on D3 for the stirrer motor.
 * A1 and A2 are analog feedback from the CS maker
*********************/

```
// include the library code:
#include <LiquidCrystal.h>

// initialize the library with the numbers of the
// interface pins
LiquidCrystal lcd(9, 8, 4, 5, 6, 7);

// Pins for Colloidal silver maker
int CS1=10;
int CS2=11;
// Pins for stirrer
int Stir=3;
int Shutdown=0;
// Variables for time
int hours;
int minutes;
int seconds;
long hour = 3600000; // 3600000 milliseconds in an hour
long minute = 60000; // 60000 milliseconds in a minute
long second = 1000; // 1000 milliseconds in a second
float AN1=0.0; // Analog inut 1
float AN2=0.0;
float temp1=0.0;
float temp2=0.0;
float CUR=0.0;   // Current in ma
float MCUR=0.0;  // Memory Current in ma
int CurSet=255;  // Current Setting

void setup() {
  // set up the LCD's number of columns and rows:
  lcd.begin(16, 2);
  pinMode (CS1, OUTPUT);
  pinMode (CS2, OUTPUT);
  pinMode (Stir, OUTPUT);
}
```

```
void loop() {
  // Reverse current every 30 minutes
  if (Shutdown==0){
    analogWrite(Stir, 128); // 1/2 supply voltage
    if (minutes<30){   // Reverse every 30 minute
      analogWrite(CS1, 0);
      analogWrite(CS2, CurSet);
    }
    else{
      analogWrite(CS2, 0);
      analogWrite(CS1, CurSet);
    }
  }
  else{  // Shutdown
    analogWrite(CS1, 0);
    analogWrite(CS2, 0);
    analogWrite(Stir, 0);
  }
  temp1=analogRead(A1);
  temp2=analogRead(A2);
  AN1=((temp1*5.0)/1024.0)*4.0; // Convert to volts
  AN2=((temp2*5.0)/1024.0)*4.0;
  CUR=abs(AN1-AN2);
  if (((CUR+MCUR)/2) > 1.0) {  // Reduce PWM, current
    CurSet--;
  }
  lcd.clear();
  lcd.setCursor(0,0);
  lcd.print("V:");
  lcd.print(AN1);
  lcd.setCursor(8,0);
  lcd.print("V:");
  lcd.print(AN2);
  lcd.setCursor(10,1);
  lcd.print("C:");
  lcd.print(CUR);
  // print the number of seconds since reset:
  long timeNow = millis();
  hours = (timeNow) / hour;
  minutes = ((timeNow) % hour) / minute ;
  seconds = (((timeNow) % hour) % minute) / second;
```

```
  lcd.setCursor(0, 1);
  lcd.print("T:");
  lcd.print(hours);
  lcd.print(":");
  lcd.print(minutes);
  lcd.print(":");
  lcd.print(seconds);

  if (hours>3){ // Time is 4 hours
    Shutdown=1;
    }
  if (((CUR+MCUR)/2)>2.0){ // Average Current over 2ma
    Shutdown=1;
    }
  MCUR=CUR;   // Save current for averaging
  delay(300);
}
```

Bibliography

A lot of the support software comes for Adafruit. Please support them by purchasing their products.

Programming Arduino
Getting Started With Sketches
By Simon Mark
Copyright 2012 by the McGraw-Hill Companies

This book gives a thorough explanation of the programming code for the Arduino. However the projects in the book are very basic.

Getting Started with Arduino
By Massimo Banzi
Copyright 2011 Massimo Banzi

This author is a co-founder of the Arduino. This book has a quick reference to the programming code and some simple projects.

Arduino Cookbook
by Michael Margolis
Copyright © 2011 Michael Margolis and Nicholas Weldin. All rights reserved.
Printed in the United States of America.
Published by O'Reilly Media, Inc., 1005 Gravenstein Highway North, Sebastopol, CA.

This book has lots of great projects, with a very good explanation for every project.

Practical Arduino: Cool Projects for Open Source Hardware
Copyright © 2009 by Jonathan Oxer and Hugh Blemings
ISBN-13 (pbk): 978-1-4302-2477-8
ISBN-13 (electronic): 978-1-4302-2478-5
Printed and bound in the United States of America

www.ingramcontent.com/pod-product-compliance
Lightning Source LLC
Chambersburg PA
CBHW062356220526
45472CB00008B/1825